ENVIRONMENTAL SCIENCE, ENGINEERING AND TECHNOLOGY

PROTECTING THE GREAT LAKES FROM INVASIVE AND NONINDIGENOUS SPECIES

Environmental Science, Engineering and Technology

Additional books in this series can be found on Nova's website under the Series tab.

Additional E-books in this series can be found on Nova's website under the E-books tab.

ENVIRONMENTAL SCIENCE, ENGINEERING AND TECHNOLOGY

PROTECTING THE GREAT LAKES FROM INVASIVE AND NONINDIGENOUS SPECIES

CLARA E. WOUTERS
EDITOR

Nova
Nova Science Publishers, Inc.
New York

For permission to use material from this book please contact us:
Telephone 631-231-7269; Fax 631-231-8175
Web Site: http://www.novapublishers.com

Library of Congress Cataloging-in-Publication Data

Protecting the Great Lakes from invasive and nonindigenous species / editor, Clara E. Wouters.
p. cm.
Includes index.
ISBN 978-1-61728-103-7 (hardcover)
1. Nonindigenous aquatic pests--Great Lakes (North America)--Simulation methods. 2. Ballast water--Management. I. Wouters, Clara E.
TD427.A68P76 2010
628.9'60977--dc22
2010015622

Published by Nova Science Publishers, Inc. ✢ New York

CONTENTS

PREFACE

Nonindigenous species (NIS) are organisms that enter an ecosystem beyond their native spatial range. The Sea Lamprey (Petromyzon marinus) was the first to enter the Great Lakes during the 1830s facilitated by the Erie and Welland canals. Since then, at least 185 other species have invaded the Great Lakes. Thirteen of these species have been labeled as invasive by causing ecological or economic harm. This book explores the data and tools that U.S. Great Lakes resource managers can use to more effectively prevent the establishment of aquatic NIS and maps the habitats of the Great Lakes most vulnerable to the entry of aquatic NIS while identifying particular NIS that have the potential to enter the U.S. Great Lakes ports.

Chapter 1 - In recent years, many people have become increasingly aware that the globalization of trade, the increased speed of travel, the massive volume of cargo shipments, and rising tourism have combined to increase the chance of accidental introductions of foreign species into the United States. Aquatic species arrive through a variety of mechanisms—unintentionally when attached to vessel hulls or carried in vessel ballast water and intentionally when imported for aquaria display, as live seafood for human consumption, or as a transplant to increase sport fishing opportunities.

The arrival of zebra mussels in the Great Lakes and their subsequent damage to city water supplies and electric utilities has focused significant attention on ballast water discharge by cargo ships as a high-risk mechanism for species invasion. New management efforts attempt to address this concern. Congress is considering legislative proposals to amend and reauthorize the Nonindigenous Aquatic Nuisance Prevention and Control Act, including specific provisions that would modify how ballast water is managed.

In late August 2009, the U.S. Coast Guard published proposed regulations to establish quantitative standards for ballast water treatment. The proposed standards would initially follow standards developed by the International Maritime Organization. In a subsequent phase, the quantitative standards would become much more stringent, given sufficient technological development to support achievement of the higher standards. The proposed Coast Guard standards would not preempt existing state ballast water management standards.

This chapter provides background on various approaches to ballast water management and reviews current ballast water management laws and programs.

With increases in the number of people traveling, the speed and methods of travel, the types and volume of trade, the ability to move living plants and animals so that more of them survive the journey, and the different modes of transport for hitch-hiking organisms, invasive species have become a global concern. Although there are many ways in which species may invade, this chapter focuses on ballast water discharge by cargo ships as one of the more significant mechanisms for biotic invasion of coastal and estuarine habitats as well as inland navigable waters.

The arrival of zebra mussels in the Great Lakes in the late 1 980s and their subsequent damage to city water supplies and electric utilities focused initial attention on ballast water as a source of invasive species. Reflecting the scope of the problem, the San Francisco Bay-Delta ecosystem is considered to be one of the most disrupted aquatic ecosystems in the United States, with colonization by more than 230 non-native species. Compounding the problem, species that invaded via other mechanisms (e.g., mitten crabs that are believed to have been an illegal seafood introduction) may be further spread through ballast water transport and vice versa (e.g., zebra mussels are spreading to new drainages via boats transported on trailers). In the Gulf of Mexico, ballast water has been implicated in the contamination of commercial oyster beds. Globally, it is estimated that more than 10,000 marine species each day may be transported across the oceans in the ballast water of cargo ships. The ctenophore *Mnemiopsis leidyi* was transported in ballast water from estuaries along the Atlantic coast of North America to the Black Sea, where it is blamed for a massive collapse of the commercial fish harvest. Around 1978, the American jackknife clam, *Ensis directus*, was introduced through ballast water in the German Bight and has spread rapidly over the North Sea coast, where it has become one of the most common bivalves, replacing many native species.

The economic, social, recreational, and ecological losses/costs attributable to aquatic invasive species are difficult to quantify. While some costs have been estimated, such as the $5 billion in damages to water pipes, boat hulls, and other hard surfaces by zebra mussels in the Great Lakes, others, such as the losses of native species and environment restoration to pre-invasion quality, are unknown. Congress is considering legislative proposals to reauthorize and amend the Nonindigenous Aquatic Nuisance Prevention and Control Act of 1990, including specific provisions that would modify how ballast water is managed.

Chapter 2 - The Great Lakes of the United States have been subjected to adverse ecological and economic impacts from nonindigenous species (NIS). Ballast water from commercial shipping is the major means by which NIS have entered the Great Lakes. To help resource managers assess the future arrival and spread of invasive species, 58 species were initially identified as having a moderate or high potential to spread and cause ecological impacts to the Great Lakes. Using a species distribution model (the Genetic Algorithm for Rule-Set Production or GARP), areas within the Great Lakes where 14 of these 58 potential invasive species could find suitable habitat, were identified. Based on the model and species depth tolerances, all of Lake Erie and the shallow water areas of the other four Great Lakes are most vulnerable to invasion by the 14 modeled species. Analysis of ballast water discharge data of vessels entering the Great Lakes via the St. Lawrence Seaway revealed that the original source of most ballast water discharges came from Canada and Western Europe. The Great Lakes ports at greatest risk for invasion by the 14 modeled species from ballast water discharges are Toledo, Ashtabula and Sandusky, OH; Gary, IN; Duluth, MN; Milwaukee and Superior, WI; and Chicago, IL. Since early detection is critical in managing for NIS, these results should help focus monitoring activities on particular species at the most vulnerable Great Lakes ports. This assessment demonstrates that successful invasions are best predicted by knowing the propagule pressure (i.e., the number of larvae/individuals entering a new area) and habitat matching (i.e., how similar is the invaded area to the native range of the species).

In: Protecting the Great Lakes from Invasive and Nonindigenous... ISBN: 978-1-61728-103-7
Editor: Clara E. Wouters

Chapter 1

BALLAST WATER MANAGEMENT TO COMBAT INVASIVE SPECIES

Eugene H. Buck

SUMMARY

In recent years, many people have become increasingly aware that the globalization of trade, the increased speed of travel, the massive volume of cargo shipments, and rising tourism have combined to increase the chance of accidental introductions of foreign species into the United States. Aquatic species arrive through a variety of mechanisms—unintentionally when attached to vessel hulls or carried in vessel ballast water and intentionally when imported for aquaria display, as live seafood for human consumption, or as a transplant to increase sport fishing opportunities.

The arrival of zebra mussels in the Great Lakes and their subsequent damage to city water supplies and electric utilities has focused significant attention on ballast water discharge by cargo ships as a high-risk mechanism for species invasion. New management efforts attempt to address this concern. Congress is considering legislative proposals to amend and reauthorize the Nonindigenous Aquatic Nuisance Prevention and Control Act, including specific provisions that would modify how ballast water is managed.

In late August 2009, the U.S. Coast Guard published proposed regulations to establish quantitative standards for ballast water treatment. The proposed standards would initially follow standards developed by the International Maritime Organization. In a subsequent phase, the quantitative standards would become much more stringent, given sufficient technological development to support achievement of the higher standards. The proposed Coast Guard standards would not preempt existing state ballast water management standards.

This chapter provides background on various approaches to ballast water management and reviews current ballast water management laws and programs.

With increases in the number of people traveling, the speed and methods of travel, the types and volume of trade, the ability to move living plants and animals so that more of them survive the journey, and the different modes of transport for hitch-hiking organisms, invasive species have become a global concern. Although there are many ways in which species may invade,[1] this chapter focuses on ballast water discharge by cargo ships as one of the more significant mechanisms for biotic invasion of coastal and estuarine habitats as well as inland navigable waters.

The arrival of zebra mussels in the Great Lakes in the late 1 980s and their subsequent damage to city water supplies and electric utilities[2] focused initial attention on ballast water as a source of invasive species. Reflecting the scope of the problem, the San Francisco Bay-Delta ecosystem is considered to be one of the most disrupted aquatic ecosystems in the United States, with colonization by more than 230 non-native species.[3] Compounding the problem, species that invaded via other mechanisms (e.g., mitten crabs that are believed to have been an illegal seafood introduction) may be further spread through ballast water transport and vice versa (e.g., zebra mussels are spreading to new drainages via boats transported on trailers). In the Gulf of Mexico, ballast water has been implicated in the contamination of commercial oyster beds.[4] Globally, it is estimated that more than 10,000 marine species each day may be transported across the oceans in the ballast water of cargo ships.[5] The ctenophore *Mnemiopsis leidyi* was transported in ballast water from estuaries along the Atlantic coast of North America to the Black Sea, where it is blamed for a massive collapse of the commercial fish harvest. Around 1978, the American jackknife clam, *Ensis directus*, was introduced through ballast water in the German Bight and has spread rapidly over the North Sea coast, where it has become one of the most common bivalves, replacing many native species.

The economic, social, recreational, and ecological losses/costs attributable to aquatic invasive species are difficult to quantify. While some costs have been estimated, such as the $5 billion in damages to water pipes, boat hulls, and other hard surfaces by zebra mussels in the Great Lakes,[6] others, such as the losses of native species and environment restoration to pre-invasion quality, are unknown. Congress is considering legislative proposals to reauthorize and amend the Nonindigenous Aquatic Nuisance Prevention and Control Act of 1990,[7] including specific provisions that would modify how ballast water is managed.

BALLAST WATER MANAGEMENT

Ballast water is held in tanks and/or cargo holds of ships to provide stability and maneuverability during a voyage when ships are not carrying cargo, are not carrying heavy enough cargo, or require more stability due to rough seas.[8] Ballast water may be either fresh or saline. Ballast water may also be carried so that a ship rides low enough in the water to pass under bridges and other structures. Ballast water management (BWM) for vessels includes all measures that aim to prevent unwanted aquatic nuisance species from being transported between ports in the ballast. Seaports in which ships exchange ballast water daily are at severe risk of invasions. Organisms transported to U.S. ports from foreign harbors with similar physiochemical characteristics (e.g., water temperatures, salinity regimes) pose an especially high risk of invasion. Even if only a tiny proportion of newly arriving non-native species survive in new habitats, such as San Francisco Bay, Chesapeake Bay, or Boston Harbor, the actual number of successful invasive species can be very large.

There are several different ways of managing ballast water. Currently, the most widely used is ballast water exchange. Ballast water exchange means that ships on their way to the next port release the lower-salinity coastal water they brought aboard in their last port and replace it with higher-salinity open-ocean water. Although this measure is not perfect, it reduces the number of potentially invasive species in the ballast tanks and replaces them with oceanic organisms that are less likely to survive in the lower-salinity near-shore waters of the ship's next port.[9] However, organisms with a wide tolerance for differing salinities may survive ballast water exchange, especially any such organisms that may reside in the unpumpable residual water and sediment remaining in the tanks during any ballast water exchange.

Another approach to BWM is treatment.[10] Ballast water treatment is the subject of extensive current research and development, and several technologies and methodologies have been proposed. These include mechanical methods (e.g., filtration and separation), physical methods (e.g., sterilization by ultraviolet light, ozone, heat, electric current, or ultrasound), and chemical methods (using biocides). In addition, treatment may combine several of these methods.[11]

Treatment may be an appropriate management option on occasions when vessels temporarily operate without ballast—a "no-ballast-on-board" (NOBOB) situation. When a ship is carrying no ballast water, it presents unique treatment problems because large numbers of organisms can reside in the unpumpable residual water and sediment remaining in the ballast tanks. Few of the tested methodologies have been applied to the control of organisms in NOBOB situations.[12] The treatment option favored by many ship operators because of its intrinsic simplicity and relatively low cost is biocide application, whereby chemical agents are added to the ballast water to minimize the number of (i.e., kill) viable organisms. This approach also has the potential to address the NOBOB condition.[13] Concerns remain relating to establishing and enforcing standards for the appropriate disposal of biocide-treated ballast water and sediments.

Although estimates of the costs of ballast treatment may be imprecise and vary from vessel to vessel, there is some general agreement on average costs.[14] For example, it may cost an estimated $400,000 per vessel for modification of container/bulk vessels to use onshore ballast water treatment facilities at California ports. More generally, the cost of retrofitting vessels to treat ballast water has been estimated at between $200,000 and $310,000 per vessel for mechanical treatment and around $300,000 for chemical treatment.[15] Most of this expense will be borne by foreign shipping companies, as the U.S. flag fleet is a small percentage of the global fleet,[16] and likely passed along to consumers of products imported on these ships. The likelihood of compliance by the foreign flag fleet was increased by the February 2004 conclusion of an international agreement on ballast water management (see *"International Efforts,"* on Page 9).

CURRENT U.S. LAW

Attempts to address ballast water concerns in the United States began with the Nonindigenous Aquatic Nuisance Prevention and Control Act of 1990 (NANPCA),[17] which established a federal program to prevent the introduction and to control the spread of unintentionally introduced aquatic nuisance species. The U.S. Coast Guard, U.S. Environmental Protection Agency (EPA), U.S. Fish and Wildlife Service (FWS), Army Corps of Engineers, and National Oceanic and Atmospheric Administration (NOAA) shared responsibilities for implementing this effort, acting cooperatively as members of an Aquatic Nuisance Species (ANS) Task Force to conduct studies and report to Congress to (1) identify areas where ballast water exchange can take place without causing environmental damage; and (2) determine the need for controls on vessels entering U.S. waters other than the Great Lakes. Under § 1101 of NANPCA, a Great Lakes BWM program (voluntary in its first two years) became mandatory in 1992. This section directed the Coast Guard to issue regulations (33 CFR Part 151, Subpart C) to prevent the introduction and spread of aquatic nuisance species into the Great Lakes through the ballast water of vessels and established civil and criminal penalties for violating these regulations. NANPCA also encouraged the Secretary of Transportation (but now the Secretary of Homeland Security) to negotiate with foreign countries, through the International Maritime Organization, to prevent and control the unintentional introduction of aquatic nuisance species.

In 1996, the National Invasive Species Act (NISA) amended NANPCA to create a national ballast management program modeled after the Great Lakes program wherein all ships entering U.S. waters (after operating outside the U.S. Exclusive Economic Zone) are directed to undertake high seas (i.e., mid-ocean) ballast exchange or alternative measures pre-approved by the Coast Guard as equally or more effective. While not initially enforced on a ship-by-ship basis, this national program was to have become mandatory within three years of the date the Coast Guard issued interim voluntary guidelines[18] if ships did not show adequate compliance with the program[19] in the absence of enforcement.

The National Ballast Information Clearinghouse (NBIC) was developed jointly by the Coast Guard and the Smithsonian Environmental Research Center to synthesize, analyze, and interpret national data concerning BWM. During the first two years (July 1999 through June 2001), the NBIC found that nationwide compliance with ballast exchange reporting requirements was low, with only 30.4% of vessels entering the U.S. Exclusive Economic Zone (EEZ) filing reports with the NBIC.[20] In addition, a Coast Guard Report to Congress concluded that reporting compliance was insufficient to allow an accurate assessment of voluntary BWM.[21] On January 6, 2003, the Coast Guard proposed penalties for those who failed to submit BWM reports required by 33 U.S.C. § 151 Subpart D for most vessels entering U.S. waters.[22] On June 14, 2004, the Coast Guard published final regulations establishing penalties for failure to comply with reporting requirements.[23] On July 28, 2004, the Coast Guard published final regulations changing the voluntary ballast water management program to a mandatory one.[24]

NISA encouraged negotiations with foreign governments to develop and implement an international program for preventing the introduction and spread of invasive species in ballast water. NISA also required a Coast Guard study and report to the Congress on the effectiveness of existing shoreside ballast water facilities used by crude oil tankers in the coastal trade off Alaska, as well as studies of Lake Champlain, the Chesapeake Bay, San Francisco Bay, Honolulu Harbor, the Columbia River system, and estuaries of national significance.[25] Under NISA, a Ballast Water Management Demonstration Program was authorized to promote the research and development of technological alternatives to ballast water exchange.

NISA has been criticized as inadequate and faulted for several alleged shortcomings, including agency weakness or delay in implementing some of its provisions.[26] Since NISA exempted most coastwise vessel traffic from ballast water exchange guidelines, vessels traveling short distances between U.S. ports (e.g., from San Francisco Bay, which is highly invaded, to Puget Sound, which is less so) are exempt from controls. Others are critical of the provisions of 16 U.S.C. §4711(k)(2)(A) giving the vessel owner a blanket exemption to ignore any mandatory regulations if the master determines that the vessel might not be able to safely conduct a ballast water exchange on the open ocean. Whereas earlier provisions applicable to the Great Lakes provided a safety exemption, the master/captain of a vessel was required to report the problem to the Coast Guard and conduct alternate BWM measures, often negotiated on a case-by-case basis. Critics believe the NISA language has eliminated any incentive to change ballast water piping systems or adopt other management or treatment options to deal with the problem safely. Finally, NISA has been criticized for its apparent failure to actually prevent additional introductions of damaging organisms into the Great Lakes, despite this being the one area where the requirements for managing ballast water have been the most stringent for the longest time.[27]

FEDERAL AGENCY PROGRAMS

U.S. Coast Guard

Under NANPCA, the Coast Guard is responsible for developing and implementing a BWM program to prevent the unintentional introduction and dispersal of nonindigenous aquatic species into waters of the United States from ship ballast water.[28] Initially this was accomplished through a mandatory BWM program for the Great Lakes ecosystem and voluntary guidelines for the remainder of U.S. waters. Relevant regulations, published at 33 CFR Part 151, Subpart C, went into effect in 1993.[29] Under these regulations, the Coast Guard enforced mandatory requirements for ballast water management only for the Great Lakes. Ballast water reporting data for inbound vessels was submitted via fax either directly to the Captain of the Port (COTP) Buffalo or U.S. Coast Guard's Marine Safety Detachment (MSD) Massena, or via the Saint Lawrence Seaway Development Corporation (SLSDC) to MSD Massena at least 24 hours before arrival (33 CFR 151.2040). Compliance with these requirements was essentially 100%. Every vessel that reports that it is carrying ballast water on board (BOB) while transiting to the Great Lakes underwent ballast water inspections by either MSD Massena at the locks in Massena, NY, or by the SLSDC in Montreal. Compliance with the mandatory ballast water requirements averaged approximately 92% in the late 1990s and early 2000s for those vessels declaring BOB. Those vessels found to not have conducted proper exchange were ordered not to discharge ballast water in the Lakes. For these vessels, or if the vessel declared that it intended to retain its ballast water on board while in the Great Lakes system, then MSD Massena again boarded the vessel after it has made port calls in the Lakes and passed the lock in Massena on its outbound transit to ensure the vessel had not discharged while in the Lakes. Vessel compliance with the order not to discharge has been excellent.[30]

As stated in the Secretary of Transportation's November 2001 *United States Coast Guard Report to Congress on the Voluntary National Guidelines for Ballast Water Management*,[31] the Coast Guard began developing regulations requiring active BWM of all ships that enter U.S. waters after operating beyond the EEZ, including sanctions for failure to comply. On July 30, 2003, the Coast Guard proposed mandatory BWM practices for all vessels with ballast tanks bound for ports or places within the United States or entering U.S. waters, but excluding domestic port-toport voyages.[32] The Coast Guard published final regulations on July 28, 2004, extending BWM requirements already in place for vessels entering the Great Lakes and Hudson River to all U.S. ports and waters.[33] The Coast Guard requested comments on potential revisions to mandatory ballast water management reporting requirements in early 2007.[34]

The Coast Guard also has taken action to establish a quantitative ballast water treatment performance standard; protocols for testing, verifying, and reporting on treatment technologies; and a program to facilitate experimental shipboard installation and operation of promising technologies. On March 4, 2002, the Coast Guard published an advance notice of proposed rulemaking, seeking comments on development of a ballast water treatment goal and an interim ballast water treatment standard as part of regulations that would make guidelines for ballast exchange mandatory.[35] On September 26, 2003, the U.S. Coast Guard announced its intent to prepare a Programmatic Environmental Impact Statement for proposed regulatory action to establish a ballast water discharge standard.[36] Such a standard would establish the required level of environmental protection necessary to prevent introductions and combat the spread of invasive species from ballast water discharges. Comments were accepted on this proposal through December 26, 2003. On May 2, 2006, the U.S. Coast Guard published a request for public comment on the status of research and development of ballast water management systems and analytical technologies for testing such systems.[37] On August 28, 2009, the Coast Guard published proposed regulations to establish standards for ballast water treatment.[38] The proposed standards would initially follow standards developed by the International Maritime Organization (see "International Efforts," below), and would become much more stringent in a subsequent phase, given sufficient technological development to support achievement of these higher standards. The proposed Coast Guard standards would not preempt state ballast water management standards (see "State Programs," below).

On January 2, 2004, the U.S. Coast Guard announced the beginning of a program to facilitate the installation of experimental shipboard ballast water treatment systems on both foreign and domestic vessels.[39] This Shipboard Technology Evaluation Program (STEP) aims to promote research and development of shipboard ballast water treatment systems through regulatory incentives, creating more options for vessel owners seeking alternatives to ballast water exchange. Regulatory incentives would grant conditional equivalencies for accepted vessels in the STEP that might not meet discharge standards mandated by future regulations. On August 5, 2004, the U.S. Coast Guard announced its interest in establishing a program to approve ballast water treatment systems to assure that approved systems meet ballast water discharge standards the Coast Guard anticipates implementing.[40] On August 28, 2009, the Coast Guard published proposed regulations that would establish an approval process for ballast water management systems.[41]

On August 31, 2005, the Coast Guard published a notice of policy establishing best management practices for NOBOB vessels entering the Great Lakes that have residual ballast water and ballast tank sediment.[42]

Department of State

The Department of State participated in negotiations through the International Maritime Organization to develop an international convention to control the spread of invasive species from the exchange of ships' ballast water. (See "International Efforts," below.)

National Oceanic and Atmospheric Administration

Through both the National Sea Grant College program and a BWM technology development program, NOAA has funded research on alternatives to ballast water exchange as methods of BWM.[43] NOAA's Great Lakes Environmental Research Lab (GLERL) targets prevention and control to stop the inflow and spread of new aquatic organisms, with particular emphasis on ship ballast. GLERL, with combined funding from NOAA and several other agencies, developed and provides leadership for the Great Lakes NOBOB and Ballast Exchange research program, focusing on the biological assessment of ballast tank residuals and the experimental determination of effectiveness of ballast exchange. In this program, GLERL scientists collaborate with scientists at several universities and the Smithsonian Environmental Research Center. In a related project, scientists at GLERL and the University of Michigan are evaluating two chemicals for use on residuals in NOBOB tanks. In addition, GLERL has been working on developing a model of ballast tank flow during ballast tank exchange. NOAA also manages the Ballast Water Management Demonstration Program grant competition.[44]

Department of Defense

The Department of Defense (DOD) is promulgating joint regulations with the EPA covering discharges from DOD vessels (40 CFR 1700) to implement §312(n) of the Clean Water Act (33 U.S.C. § 1322(n)). When complete, they will set discharge standards for vessel ballast water to address the environmental effect of non-native species introduction via that ballast water (as well as addressing chemical pollution from other Armed Forces vessel discharges). The regulations are being developed in three phases. The first, completed in May 1999, determined which ballast water discharges would require control. The second, currently in progress, will set performance standards, and the third will promulgate regulations for meeting those standards.

Smithsonian Institution

The Smithsonian Environmental Research Center performs research to examine patterns of ballast water delivery and measures species transfer associated with shipping. In cooperation with the Coast Guard, the center established the National Ballast Information Clearinghouse (NBIC)[45] to measure the changing patterns of ballast water delivery and management for vessels arriving in U.S. ports and to synthesize national data on patterns and impacts of alien species in coastal ecosystems.

Environmental Protection Agency

The EPA establishes water quality standards and regulates discharges under the Clean Water Act.[46] On September 9, 2003, the EPA announced that it was denying a 1999 petition by the Pacific Environmental Advocacy Center, Center for Marine Conservation, San Francisco Bay Keeper, and a number of other concerned groups to require regulation of vessel ballast water discharges through Clean Water Act permits, arguing that the Coast Guard was the more appropriate regulatory agency.[47] This decision did not alter the development and implementation of the joint EPA-DOD regulations covering discharges from DOD vessels (discussed above) because this DOD effort is specifically authorized in statute. In late December 2003, three Pacific Coast environmental groups filed suit, seeking to force the EPA to regulate ballast water discharges under the Clean Water Act.[48] On September 18, 2006, the federal district court ruled that EPA's regulation exempting ballast water discharges from the Clean Water Act was contrary to congressional intent and ordered EPA to promulgate new regulations within two years (i.e., September 30, 2008).[49] This ruling, which the government has appealed to the U.S. Court of Appeals for the Ninth Circuit, directs EPA to ensure that shipping companies and other vessels comply with the Clean Water Act by restricting the discharge of invasive species in ballast water under that act's permit program. In response, EPA published regulations for Vessel General Permits on December 29, 2008.[50] Environmental groups have expressed concern over the effectiveness of the EPA regulations.[51]

In light of this ruling, concern arose in the 110th Congress as to what extent legislation proposing national standards for ballast water management to be administered by the Coast Guard, such as those outlined in H.R. 2830 and S. 1578, might conflict with EPA regulation of ballast water discharges as water pollutants under the Clean Water Act.[52] The 110th Congress enacted P.L. 110- 299 to provide a two-year moratorium on Clean Water Act permitting for discharges (except for ballast water discharges) from commercial fishing vessels and non-recreational vessels less than 79 feet in length in response to the Northwest Environmental Advocates litigation.[53] During the moratorium, EPA is to study the discharges from these vessels and submit a report to Congress.

State Programs

In addition to federal programs and sometimes in response to perceived deficiencies in federal regulation, several states have chosen to regulate aspects of ballast water management, including Maryland,[54] California,[55] Oregon,[56] Washington,[57] and Michigan.[58] In light of this state action, provisions in some of the measures that were considered in the 110th Congress, such as H.R. 2830 and S. 1578, would have preempted state ballast water discharge standards that are more stringent than federal standards, while continuing to allow states to inspect and enforce federal ballast water standards.

International Efforts

In July 1991, the International Maritime Organization's (IMO 's) Marine Environmental Protection Committee (MEPC) issued voluntary *International Guidelines for Preventing the Introduction of Unwanted Aquatic Organisms and Pathogens for Ships'Ballast Water and Sediment Discharges*, adopted in Resolution (50) 31 by a diplomatic conference of the IMO in November 1993. IMO members were requested to follow these guidelines, which also called for exchange of ballast water in the open ocean (to reduce transfer of species from port to port). A review conducted by Australia in 1993 revealed that few countries had implemented the guidelines. In 1994, the MEPC established a ballast water working group to draft regulations for the control and management of ships' ballast water. These draft regulations were debated at the November 1998 and June 1999 MEPC meetings. IMO subsequently proposed these management protocols as a formal IMO instrument. This instrument requires all ratifying member nations to follow the regulations, which would include open-ocean exchange.

On February 13, 2004, an *International Convention for the Control and Management of Ships' Ballast Water and Sediments* was adopted at the International Conference on Ballast Water Management for Ships in London, England. This Convention will enter into force 12 months after ratification by 30 nations, representing 35% of the world merchant shipping tonnage.[59] The United States was one of the major proponents of this convention, which requires all ships to implement a Ballast Water and Sediments Management Plan. In addition, all ships must carry a Ballast Water Record Book and will be required to conduct BWM procedures in conformity with a specified standard. These standards will be phased in for various vessels, depending upon when they were constructed.[60] Resistance by vessel owners to the new international requirements for ballast water treatment is tempered by their interest in global standards under the IMO Convention as opposed to the increasing number of national ballast water programs developed around different approaches for addressing this concern. However, U.S. legislation under consideration by the 111th Congress could mandate ballast water treatment standards as much as 100 times more stringent than those required by the IMO Convention, possibly making the IMO Convention irrelevant, if the more stringent standards could be met with readily available treatment technology at no substantially greater expense.[61] A Global Industry Alliance—a partnership among IMO, the United Nations Development Program, the Global Environment Facility, and four private shipping corporations—was formed in March 2009 to coordinate international research and development of cost-effective BWM technologies.[62]

Bilaterally with Canada, the United States is cooperating through the North American Commission for Environmental Cooperation (CEC),[63] the Great Lakes Commission, and the International Joint Commission to better understand, coordinate, and address ballast water management concerns. Improved ballast water management is one component of the CEC's project on "Closing the Pathways of Aquatic Invasive Species across North America."[64] The Great Lakes Commission convened the "Great Lakes Panel on Aquatic Nuisance Species" in late 1991 to coordinate efforts, including ballast water management.[65] In addition, the International Joint Commission (IJC) views invasive species management as an important component of water quality.[66] Furthermore, the Saint Lawrence Seaway Development Corporation jointly administers regulations and seeks to harmonize ballast water regulations for vessels transiting U.S. waters of the Seaway after having operated outside the U.S. EEZ with those required by Canadian authorities for transit in waters under Canadian jurisdiction.[67]

Congressional Action

Numerous congressional hearings on ballast water issues and implementation of NISA were held in recent Congresses.[68] Certain testimony at these hearings has been particularly critical of the slow progress in promulgating regulations to implement NISA. Information on legislative action to address various aspects of BWM issues in the 110th Congress can be found in CRS Report RL34640, *Regulating Ballast Water Discharges: Legislative Issues in the 110th Congress*, by Claudia Copeland. Information on bills introduced and legislative action to address various aspects of BWM issues in the 111th Congress can be found in CRS Report R40 172, *Fishery, Aquaculture, and Marine Mammal Issues in the 111th Congress*, by Eugene H. Buck and Harold F. Upton, under the "Invasive Species" heading.

End Notes

1 For an in-depth discussion of invasive species generally and their various mechanisms of invasion, see CRS Report RL30 123, *Invasive Non-Native Species: Background and Issues for Congress*, by M. Lynne Corn et al. Zebra mussels and other species arriving in ballast water are specifically discussed in "A Gallery of Harmful Non-Native Plants and Animals" in that report.

2 Colonies of zebra mussels accumulate and block water-intake pipes and screens of drinking water facilities, industrial facilities, power-generating plants, golf course irrigation pipes, and cooling systems of boat engines.

3 Information from the Natural Resources Defense Council webpage at http://www.nrdc.org/greengate/wildlife invasivef.asp.

4 Fred C. Dobbs and Andrew Rogerson, "Ridding Ships' Ballast Water of Microorganisms," *Environmental Science & Technology*, v. 39, no. 12 (June 15, 2005): 259A-264A.

5 G. Tracy Mehan, Assistant Administrator for Water, U.S. Environmental Protection Agency, testimony before the Senate Committee on Environment and Public Works, Subcommittee on Fisheries, Wildlife, and Water, June 17, 2003.

6 16 U.S.C. §4701(a)(4).

7 Title I of P.L. 101-646; 16 U.S.C. §§4701-4741.

8 Massachusetts Institute of Technology Sea Grant, *Marine Bioinvasions Fact Sheet: Ballast Water*, Cambridge, MA, Dec. 4, 2002; available at http://massbay.mit.edu/exoticspecies/ballast/fact.html.

[9] National Ballast Information Clearinghouse, *Present Ballast Water Management Practices*; available at http://invasions.si.edu/nbic/managementpract.html.

[10] A detailed discussion and evaluation of various ballast water treatment options is presented in Chapter 4 ("Shipboard Treatment Options") of Commission on Engineering and Technical Systems, *Stemming the Tide: Controlling Introductions of Nonindigenous Species by Ships' Ballast Water*, National Academy of Sciences (Washington, DC: 1996), 141 pp.

[11] For example, Australian researchers established a pilot plant to sterilize ballast water using a combination of filtration, ultraviolet light, and sonic disintegration. For more information, see David Salt, "Shipboard Pests get Sterile Treatment," *ABC Science Online*, Sept. 30, 2003; available at http://www.abc.net.au /science/news/enviro/ EnviroRepublish_956770.htm.

[12] The Coast Guard is studying methods to address NOBOB concerns; for more information, see 70 *Fed. Reg.* 1448-1449 (Jan. 7, 2005).

[13] David T. Stocks, "Biocides Testing in the Great Lakes," *BMT Focus* (Summer 2002): 12; available at http://www.bmt.org/brochures/Focus%20Summer%202002.pdf.

[14] State Water Resources Control Board, *Evaluation of Ballast Water Treatment Technology for Control of Nonindigenous Aquatic Organisms*, California Environmental Protection Agency (December 2002); available at http://www.calepa.ca.gov/Publications/Reports/Mandated/2002/BallastWater.pdf.

[15] Frans J. Tjallingii, *Market Opportunities for Ballast Water Treatment*, Royal Haskoning, International Ballast Technology Investment Fair, Chicago, IL, Sept. 21, 2001, available at http://www.nemw.org/fairtjallingii.pdf.

[16] As of July 2003, U.S. flag vessels comprised 1.65% of the global merchant fleet tonnage, according to statistics from U.S. Maritime Administration, available at http://www.marad.dot.gov/Marad_Statistics/mfw-7-03.htm.

[17] Title I of P.L. 101-646; 16 U.S.C. §§4701, et seq.

[18] 64 *Fed. Reg.* 26672-26690 (May 17, 1999). The voluntary guideline interim regulations were effective July 1, 1999. Final regulations were published in 66 *Fed. Reg.* 58381-58393 (Nov. 21, 2001).

[19] If the voluntary program did not result in sufficient compliance, reporting of BWM practices would become mandatory for nearly all vessels entering U.S. waters (33 CFR 151.2040).

[20] G. M. Ruiz, et al., *Status and Trends of Ballast Water Management in the United States: First Biennial Report of the National Ballast Information Clearinghouse* (Edgewater, MD: Smithsonian Environmental Research Center, Nov. 16, 2001), p. 4.

[21] U.S. Coast Guard, *Report to Congress on the Effectiveness of the Voluntary B WM Program*, June 3, 2002.

[22] 68 *Fed. Reg.* 523-530 (Jan. 6, 2003).

[23] 69 *Fed. Reg.* 32864-32871 (June 14, 2004).

[24] 69 *Fed. Reg.* 44952-44961 (July 28, 2004).

[25] As defined for the National Estuary Program in 33 U.S.C. § 1330.

[26] Letter of Feb. 11, 1999, to Hon. Carol Browner, Administrator of the U.S. Environmental Protection Agency, from Representatives George Miller, Jim Saxton, and 16 other Members of the U.S. House of Representatives.

[27] Union of Concerned Scientists, *The National Invasive Species Act: An Information Update* (Cambridge, MA: August 2002); available at http://www.ucsusa.org/assets

[28] An April 29, 2008 lawsuit by the Izaak Walton League and several other environmental groups in a Minnesota federal district court seeks to require the Coast Guard to prevent vessels from taking on and moving ballast water from waters of the Great Lakes to prevent the spread of viral hemorrhagic septicemia to Lake Superior from adjacent lakes. For additional information, see http://www.greatlakeslaw.org/blog

[29] 58 *Fed. Reg.* 18330-18335 (Apr. 8, 1993). Regulations were extended to include vessels entering the Hudson River in 59 *Fed. Reg.* 67632-67634 (Dec. 30, 1994).

[30] Personal communication with Jill R. Teixeira, Lieutenant Junior Grade, U.S. Coast Guard, Office of Congressional & Governmental Affairs, Washington, DC, Mar. 30, 2004.

[31] This chapter is available at http://dmses.dot.gov/docimages/pdf84/211672_web

[32] 68 *Fed. Reg.* 44691-44696 (July 30, 2003).

[33] 69 *Fed. Reg.* 44952-44961 (July 28, 2004).

[34] 72 *Fed. Reg.* 2536-2537 (Jan. 19, 2007).

[35] 67 *Fed. Reg.* 9632-963 8 (Mar. 4, 2002).

[36] 68 *Fed. Reg.* 55559-55563 (Sept. 26, 2003).

[37] 71 *Fed. Reg.* 25798-25799 (May 2, 2006).

[38] 74 *Fed. Reg.* 44632-44672 (Aug. 28, 2009).

[39] For additional information, see http://www.uscg.mil/hq/g-m/nvic/04/NVIC_01-04.pdf.

[40] 69 *Fed. Reg.* 47453-47454 (Aug. 5, 2004).

[41] 74 *Fed. Reg.* 44632-44672 (Aug. 28, 2009).

[42] 70 *Fed. Reg.* 5183 1-51836 (Aug. 31, 2005).

[43] The most recent request for proposals for ballast water treatment technology testing and demonstration projects was published by NOAA at 71 *Fed. Reg.*33898-33929 (June 12, 2006).

[44] A notice of grant funding available for FY2008 was published at 72 *Fed. Reg.* 73325-73335 (Dec. 27, 2007).

[45] For more information on the NBIC, see http://invasions.si.edu/NBIC/ballast.html.

[46] 33 U.S.C. §§1251, et seq.

[47] 68 *Fed. Reg.* 53165-53166 (Sept. 9, 2003). A copy of the EPA decision was available at http://www.epa.gov/npdes/ pubs/ballast_report_petition_response.pdf.

[48] Cathy Zollo, "Environmental Groups Seek EPA Rules for Ballast Water," *Naples Daily News*, Dec. 24, 2003, available at http://www.naplesnews.com/news/2003/dec/24/ndn_environmental_groups

[49] *Northwest Environmental Advocates, et al. v. U.S. Environmental Protection Agency*, Case No. C 03-05760 SI (N.D. Cal., decided Sept. 18, 2006).

[50] 73 *Fed. Reg.* 79473-79481.

[51] For example, see http://www.cbbulletin.com/314010.aspx.

[52] Detailed information on legislation addressing BWM issues in the 110th Congress can be found in CRS Report RL34640, *Regulating Ballast Water Discharges: Legislative Issues in the 110th Congress*, by Claudia Copeland.

[53] See CRS Report RS22878, *Clean Water Act: 110th Congress Legislation on Discharges from Recreational Boats*.

[54] See http://www.mde.state.md.us/businessinfocenter/pollutionprevention/ballast%20water% 20management/ index.asp for details on Maryland's program.

[55] See http://www.slc.ca.gov/division 40%20appendix%20e.doc for details on California's program.

[56] See http://www.deq.state.or.us/wmc/pubs/factsheets/cu/oregonballastwatermanagement.pdf for details on Oregon's program.

[57] See http://www.wdfw.wa.gov/fish for details on Washington's program.

[58] See http://www.olemiss.edu/orgs/SGLC/National/Ballast%20Water.pdf for details on Michigan's program.

[59] As of May 31, 2008, this Convention had been ratified by 14 nations, representing 3.55% of the world merchant shipping tonnage. Updated status information can be found at http://www.imo.org/Conventions mainframe.asp? topic_id=247.

[60] Details on this new Convention were available at http://globallast.imo.org/index.asp?page=mepc.htm&menu=true.

[61] Rajesh Joshi, "New US Ballast Water Plan a Threat to IMO Convention," *Lloyd's List*, May 1, 2008, available at http://www.lloydslist.com/ll/news/new-us-ballast-water-plan-a-threat

[62] IMO, *United Nations Agencies Partner with Shipping Corporations to Minimize Global Environment Threat*, IMO/UNDP Joint Press Release, March 2, 2009.

[63] Established under the authority of the North American Free Trade Agreement.

[64] Additional information on this project was available at http://www.cec.org/programs_projects/conserv_biodiv/ project/index.cfm?projectID=20&varlan=english.

[65] Additional information on this panel was available at http://www.glc.org/ans/.

[66] Additional information on these concerns was available from the IJC's annual report, available at http://www.ijc.org/ php/publications/html/1 1br/english/report/chapter3/.

[67] 73 *Fed. Reg.* 9950-9954 (Feb. 25, 2008).

[68] U.S. Congress, Committee on Resources, Subcommittee on Fisheries Conservation, Wildlife, and Oceans and Committee on Science, Subcommittee on Environment, Technology, and Standards, *H.R. 5395 and H.R. 5396*, Joint legislative hearing, 107th Cong., 2d sess., (Washington, DC: Nov. 14, 2002), 80 pp.; U.S. Congress, Committee on Resources, Subcommittee on Fisheries Conservation, Wildlife, and Oceans, and Subcommittee on National Parks, Recreation, and Public Lands, *The Growing Problem of Invasive Species*, joint legislative hearing, 108th Cong., 1st sess. (Washington, DC: Apr. 29, 2003), 133 pp.; U.S. Congress, Committee on Environment and Public Works, Subcommittee on Fisheries, Wildlife, and Water, *National Aquatic Invasive Species Act of 2003*, legislative hearing, 108th Cong, 1st sess. (Washington, DC: June 17, 2003), 240 pp.; an as-yet-unpublished joint hearing of Mar. 25, 2004, by the House Committee on Transportation and Infrastructure, Subcommittees on Coast Guard and Maritime Transportation and on Water Resources and Environment, on Ballast Water Management: New International Standards and National Invasive Species Act Reauthorization; U.S. Congress, Committee on Government Reform, Subcommittee on Regulatory Affairs, *Protecting Our Great Lakes: Ballast Water and the Impact of Invasive Species*, oversight field hearing, 109th Congress, 1st sess. (Fair Haven, MI: Sept. 9, 2005), 133 pp.; an as-yet-unpublished legislative hearing of June 15, 2005, by the Senate Committee on Commerce, Science, and Transportation, National Ocean Policy Study, on Ballast Water Invasive Species Management and Threats to Coral Reefs; U.S. Congress, Committee on Transportation and Infrastructure, Subcommittee on Water Resources and Environment, *The Impact of Aquatic Invasive Species on the Great Lakes*, legislative hearing, 1 10th Congress, 1st sess. (Washington, DC: Mar. 7, 2007), 256 pp.

In: Protecting the Great Lakes from Invasive and Nonindigenous... ISBN: 978-1-61728-103-7
Editor: Clara E. Wouters

Chapter 2

PREDICTING FUTURE INTRODUCTIONS OF NONINDIGENOUS SPECIES TO THE GREAT LAKES

National Center for Environmental Assessment

ABSTRACT

The Great Lakes of the United States have been subjected to adverse ecological and economic impacts from nonindigenous species (NIS). Ballast water from commercial shipping is the major means by which NIS have entered the Great Lakes. To help resource managers assess the future arrival and spread of invasive species, 58 species were initially identified as having a moderate or high potential to spread and cause ecological impacts to the Great Lakes. Using a species distribution model (the Genetic Algorithm for Rule-Set Production or GARP), areas within the Great Lakes where 14 of these 58 potential invasive species could find suitable habitat, were identified. Based on the model and species depth tolerances, all of Lake Erie and the shallow water areas of the other four Great Lakes are most vulnerable to invasion by the 14 modeled species. Analysis of ballast water discharge data of vessels entering the Great Lakes via the St. Lawrence Seaway revealed that the original source of most ballast water discharges came from Canada and Western Europe. The Great Lakes ports at greatest risk for invasion by the 14 modeled species from ballast water discharges are Toledo, Ashtabula and Sandusky, OH; Gary, IN; Duluth, MN; Milwaukee and Superior, WI; and Chicago, IL. Since early detection is critical in managing for NIS, these results should help focus monitoring activities on particular species at the most vulnerable Great Lakes ports. This assessment demonstrates that successful invasions are best predicted by knowing the propagule pressure (i.e., the number of larvae/individuals entering a new area) and habitat matching (i.e., how similar is the invaded area to the native range of the species).

List of Abbreviations and Acronyms

AVHRR	Advanced Very High Resolution Radiometer
BOB	ballast on board
BWD	ballast water discharge(s)
BWE	ballast water exchange
CFR	Code of Federal Regulations
EPA	Environmental Protection Agency
GARP	Genetic Algorithm for Rule-Set Production
GBIF	Global Biodiversity Information Facility
GIS	geographic information systems
K490	diffuse attenuation coefficient at 490 nm
MMT	mean monthly water surface temperature
MODIS	Moderate Resolution Imaging Spectroradiometer
NBIC	National Ballast Information Clearinghouse
NIS	nonindigenous species
NOAA	National Oceanic and Atmospheric Administration
NOBOB	no ballast on board
NOBOB-RM	no ballast on board but the vessel contains residual material
NVMC	National Vessel Movement Center
nLW	normalized water-leaving radiance
ppt	parts per thousand
sr	steradian (units of a solid angle and can also be called a squared radian)
USCG	United StatesCoast Guard
USGLP	U.S. Great Lakes ports (or ports of call)

Acknowledgments

The authors thank James Andreasen and Dan Kluza for the idea of applying the Genetic Algorithm for Rule-Set Production (GARP) model, originally developed for predicting movements of endangered species in terrestrial environments, to aquatic systems; Michael Slimak for the suggestion to apply GARP to the Great Lakes; and Anne Sergeant who served as Project Officer from 2005–2006 and provided technical direction for the literature review and initial species distribution models. The authors also thank Richard Everett, Stuart White, Bivan Patnaik, and Keith Donohue of the U. S. Coast Guard (USCG) for explaining USCG regulations, helping us interpret the National Ballast Information Clearinghouse database, and enabling their contractor, Whitman Miller of Smithsonian Environmental Research Center, to download and further explain the database.

1. Executive Summary

Nonindigenous species (NIS) are organisms that enter an ecosystem beyond their native spatial range. The Sea Lamprey (*Petromyzon marinus*) was the first to enter the Great Lakes during the 1 830s facilitated by the Erie and Welland canals. Since then, at least 185 other species have invaded the Great Lakes. Thirteen of these species have been labeled as invasive by causing ecological or economic harm. The zebra mussel (*Dreissena polymorpha*), for example, has impacted many Great Lakes native species and has imposed large expenses on the utility industry by clogging water intake pipes.

The objective of this chapter is to develop data and tools that U.S. Great Lakes resource managers can use to more effectively prevent the establishment of aquatic NIS. This study maps the habitats of the Great Lakes most vulnerable to the entry of aquatic NIS and identifies particular NIS with the potential to enter U.S. Great Lakes ports (USGLP).

Since the St. Lawrence Seaway opened in 1959, ballast water released from transoceanic vessels during commercial shipping operations has been identified as the predominant pathway for NIS to enter the Great Lakes. Transport of NIS occurs when a vessel takes-on ballast water containing NIS from outside the Seaway, the species survives in a ballast tank during transit, and is released when the ballast water is discharged into the Great Lakes. To become established in the new environment, the organisms must be able to survive, reproduce, and spread. To predict future invasions of NIS in the Great Lakes, the two most important determinants of successful invasions were evaluated: whether there is suitable habitat in the Great Lakes for nonnative species and whether there are a sufficient number of these organisms and their larvae arriving in the Great Lakes. First, a species distribution model was used to identify the areas of the Great Lakes which could provide suitable habitat for NIS of concern. Second, commercial shipping and ballast water discharge data were used to evaluate if there are a sufficient number of these organisms entering the Great Lakes to become established.

Based on a literature review of NIS life-history characteristics and invasion histories, 58 species that pose high or medium risk for becoming established in the Great Lakes and for causing ecological harm were identified. To predict the possible distributions of each of these species within the Great Lakes, spatial data sets that characterize aquatic conditions on a global scale were analyzed. These data sets were derived from remotely sensed space-based platforms, operated and made available by the National Aeronautics and Space Administration and the National Oceanic and Atmospheric Administration. Six of these data sets, each at a 4.6 km (21 km^2) spatial resolution, were found to be useful for NIS modeling. Three of the environmental variables are direct measures of water temperature and the other three indirectly relate to primary productivity and water clarity, indicators of habitat suitability.

The Genetic Algorithm for Rule-Set Production (GARP) model was used to determine habitat suitability. GARP predicts the potential distribution of species by comparing the environmental conditions of locations currently inhabited by the species (the reference area) with the environmental conditions in the region of concern. Adequate spatial distribution data were available to model only 9 of the 58 potential invasive species because the GA RP model requires at least 30 spatially unique latitude and longitude points that describe the distribution of a particular species. In addition to these nine species, GARP was also used to predict regions within the Great Lakes that would provide suitable habitat for five species of concern that were selected by the U.S. Environmental Protection Agency's Great Lakes National Program Office. Since the existing location of these five species was already known, the model was validated by comparing the reported locations of three of these species with the predicted locations. Results from GARP modeling were used to produce 14 range maps, one for each of the modeled species, predicting their locations of suitable habitat within the Great Lakes. The overall results varied with each modeled species, but generally showed that all of Lake Erie and the shallower portions of the other Great Lakes appear to be most vulnerable for invasion by the 14 modeled species. Water depth appears to be the predominant factor limiting the potential spread of many of the modeled species. Yet, at least one species, the quagga mussel (*Dreissena bugensis*), is surviving at greater depths in the Great Lakes than in its native habitat.

Releases of ballast water into USGLP were analyzed using 2006–2007 data obtained from the National Ballast Information Clearinghouse. The ports that received the most ballast water discharges from vessels entering the Seaway with ballast on board (BOB), after ballast water exchange outside the Seaway, are Duluth, MN; Toledo, OH; Superior, Green Bay, and Milwaukee, WI; and Gary, IN. The most frequent original sources of ballast water came from Antwerpen, Belgium; Puerto Cabello, Venezuela; Haraholmen, Sweden; and Bremen, Germany. It is important to note that there were no clear relationships between foreign and USGLP relative to ballast water uptake and releases. For instance, 13 vessels that discharged ballast water in Toledo obtained ballast water from 12 different foreign ports.

Some vessels enter the St. Lawrence Seaway without ballast water, but may still contain residual water or sediment containing NIS in their ballast tanks, and are referred to as no ballast on board vessels containing residual material (NOBOB-RM). After entering the Seaway, these vessels can off-load cargo and pick up ballast water which would mix with the residual material and be subsequently released into Great Lakes ports. There were considerably more discharges into USGLP from NOBOB-RM vessels than from those vessels with ballast on-board. Those ports receiving the most ballast water from NOBOB-RM vessels are Toledo, Ashtabula, and Sandusky, OH; Superior, WI; and Duluth, MN. Assuming the observed results for 2006 and 2007 are representative of discharge and shipping patterns over the past several years, the port of greatest concern for receiving sufficient propagules and providing the most suitable habitat is Toledo, OH. Toledo is located on Lake Erie, a region that the GARP model predicted would have a high chance of providing suitable habitat for the modeled species. Other ports of concern for receiving sufficient propagules and offering suitable habitat are Gary, IN; Ashtabula and Sandusky, OH; Milwaukee, WI; and Chicago, IL. Duluth, MN and Superior, WI, with high transport potential but low habitat suitability, could be a source of interlake transport of NIS.

This study involved numerous assumptions resulting in uncertain findings. A major source of uncertainty for the GARP modeling is the lack of complete occurrence data for many of the modeled species, for many parts of the globe. Another source of uncertainty is due to the lack of an ideal suite of data for characterizing aquatic environments. Data on abiotic factors such as salinity, bathymetry, substrate, pH, and nutrient levels were not available globally at the 21 km^2 scale. The lack of species-specific data on significant biotic factors, such as competition and predation, also lead to uncertainty in the modeling results. Despite these limitations, a model validation exercise confirmed that GARP and the environmental variables used could produce useful predictions of potential NIS distributions.

These predictions were validated using occurrence data from other regions to develop models that predicted known occurrences of three NIS already widespread in the Great Lakes.

There were also limitations with the vessel traffic and ballast water discharge analyses, used as a surrogate for measuring propagule pressure. First, the analysis was only based on 2 years of data, 2006–2007. A second source of uncertainty is due to the self-reporting nature of data entered into the National Ballast Information Clearinghouse. Self-reporting by vessels is not guaranteed to be accurate or complete records of actual vessel practices and should be used with caution. The analysis of discharges from NOBOB-RM vessels is also uncertain because the source of the residual material cannot be known for certain and could even be from ports within the Great Lakes. Further, this data set only includes information on the last five ports of call and species could remain in ballast tanks from visits to previous ports.

Both Canada and the United States implemented ballast water exchange procedures in 1989 and 1993, respectively. Although new NIS continue to be detected in the Great Lakes, it is possible the NIS were transported prior to 1993 and took several years to detect. Despite these procedures and subsequent regulations, it is likely that nonindigenous species will continue to arrive in the Great Lakes.

These findings support the need for detection and monitoring efforts at those ports believed to be at greatest risk. This study also demonstrates the importance of understanding invasion biology by evaluating the two most important predictors of invasion: propagule pressure and suitable habitat. Further, this may be the first time that remote sensing data were used in conjunction with GARP to predict the spread of aquatic invasive species.

2. Introduction—Nonindigenous Species Pose a Threat to Lake Ecosystems

The U.S. Great Lakes have suffered ecological damage and economic costs from a number of aquatic nonindigenous species (N IS) that have successfully invaded this region (Mills et al., 1994; NOAA, 2007a). NIS that enter an ecosystem beyond their native spatial range are expected to continue to enter the Great Lakes (Ricciardi, 2006). Preventing the transport of NIS to the region is the best way to avoid their potential adverse impacts, but if this is not possible, the next best alternative is to monitor for their arrival and control their spread. Resource managers are most concerned with NIS that may become invasive. Invasive species are nonindigenous species that are likely to cause economic or environmental harm or harm to animal or human health losses, ecological impacts, or adversely affect human health (National Invasive Species Council, 2007). Our primary goal is to help scientists and managers to better focus aquatic NIS monitoring activities and resources by identifying new invasive species, their potential to spread, and the U.S. Great Lakes ports (USGLP) most susceptible to invasion. Another goal is to demonstrate the use of a habitat suitability model and ballast water discharge data to predict invasion potential. Clients for this chapter include the U.S. Environmental Protection Agency's (EPA's) Great Lakes National Program Office, Great Lakes port officials, the U.S. Coast Guard (USCG), environmental organizations, agencies in the United States and Canada concerned about invasive species, and invasion biologists.

Our findings are intended to improve detection and monitoring programs by providing managers with an approach for (1) identifying newly established populations of invasive species, (2) tracking or detecting spatial-range expansions, and (3) estimating potential impacts of introductions or spread by gathering baseline data on pre-existing populations and habitat.

Nonindigenous species are one of the greatest threats to the world's ecosystems (Elton, 1958), and represent the greatest threat to biodiversity in lakes worldwide (Sala et al., 2000). Nonindigenous invasive species are the second most important threat to threatened and endangered species in the United States, after habitat loss or alteration (Wilcove et al., 1998). To date, about 50,000 species have been introduced into the United States (Pimentel et al., 2000). While many beneficial food crops, such as corn, wheat, and rice are included in this number, about 4,500 introduced species are free-ranging and up to one-fifth of these are invasive (U.S. Congress, 1993) and cause economic losses, ecological impacts, or adversely affect human health. The economic cost of invasive species to the United States has been estimated at $97 billion (U.S. Congress, 1993) and $137 billion (Pimentel et al., 2000) annually. Crop weeds and crop plant pathogens are the most costly ($26 and $21 billion, respectively) followed by rats and cats ($19 and $17 billion, respectively). Pimentel (2005) estimated the total environmental and economic impact (damage and control costs) of biological invaders to the Great Lakes Basin at $5.7 billion per year.

Biological invasion occurs when an organisms arrives somewhere beyond its previous range. Currently, most invasions are a result of human actions, deliberate or accidental. Fortunately, most invaders do not become pests, or reach invasive levels, but predicting those that do is difficult, at best. Invasions and introductions have long fascinated biologists from a theoretical perspective. As the economic consequences of invaders has increased, however, this fascination must now be used to develop tools that will allow one to predict future invaders, especially those that may affect whole ecosystems, such as the Great Lakes.

2.1. Nonindigenous Species and the Great Lakes

The Great Lakes have been subjected to biological invasions since the 1830s, when the sea lamprey (*Petromyzon marinus)* became the first recorded species to enter the Great Lakes from the Atlantic Ocean (Mills et al. 1993). Ricciardi (2006) reports that 182 NIS are now established in the Great Lakes and the National Oceanic and Atmospheric Administration (NOAA) reports a similar number of 185 (NOAA, 2007a; Appendix A). While any NIS may cause alterations to ecosystem structure or function, 13 of the reported NIS have become invasive (Mills et al., 1994). The zebra mussel (*Dreissena polymorpha)*, illustrates the impact of an invasive NIS. The zebra mussel is out-competing *Diporeia*, a deep-water macroinvertebrate, for food (IAGLR, 2002). *Diporeia* is a key source of food for many Great Lakes fish and has been a dominant benthic organism since the Great Lakes were formed (IAGRLR, 2002). The loss of *Diporeia* from the Great Lakes system affects the structure and function of the food web and commercially important fish such as the lake whitefish (IAGLR, 2002). Zebra mussels also appear to be responsible for more frequent occurrences of toxic algal blooms (*Microcystis*) by selectively rejecting blue-green algae as food and removing competing algae (Vanderploeg et al., 2001). From an economic standpoint, dense populations of zebra mussels have clogged water intake pipes, imposing large costs on utilities.

The St. Lawrence Seaway, which opened in 1959, is a system of canals and locks that permit ocean-going vessels (as large as 225.6 m long, 23.8 m wide, and 7.9 m deep), to travel from the Atlantic Ocean to the Great Lakes. While shipping pathways to the Great Lakes existed prior to 1959, the opening of the Seaway and technological changes in commercial shipping drastically increased international trade. The opening of the Seaway resulted in an increase in the number of ships entering the Great Lakes (Sala et al., 2000; MacIsaac et al., 2001; Duggan et al., 2003), larger ships conveying larger volumes of ballast water, and ships that have plied the waters in many geographic locations distant from the Great Lakes (Grigorovich et al., 2003a; Holeck et al., 2004; Drake et al., 2005; Duggan et al., 2005). While the transportation of goods has been economically beneficial, the unintended side effect of increased international trade has resulted in the long-range transport of NIS to the Great Lakes.

2.1.1. Origin and Patterns of Species Invasions

Most NIS that have become established in the Great Lakes since 1985 are native to the Ponto-Caspian region or the Black, Azov, and Caspian Seas (Ricciardi and MacIsaac, 2000; Appendix A). The Baltic Sea has also served as the source of many invaders in part because it has a climate very similar to the Great Lakes (Leppakoski et al., 2002) The NIS from these seas include a diverse array of taxa including fish: the round goby (*Neogobius melanostomus*), the tubenose goby (*Proterorhinus marmoratus*), the rudd (*Scardinius erythrophthalmus*), the zebra mussel, the quagga mussel (*Dreissena bugensis*), and several cladocerans (e.g., the fishhook water flea, *Cercopagis pengoi*), amphipods (e.g., *Echinogammarus ischnus*), and harpacticoid copepods (e.g., *Nitocra incerta*; MacIssaac et al., 2001; NOAA, 2007a). The success of Ponto-Caspian species may be related to their ability to survive ballast water exchange due to a broader salinity tolerance developed through a geological history that includes fluctuating water levels and salinities (Dumont, 1998).

There is no direct (i.e., nonstop) shipping traffic between the Great Lakes and the Ponto-Caspian Sea (Colautti et al., 2003), implying that the dominance of the Ponto-Caspian region as a source of invaders might be due to indirect linkages. To account for invasions where no direct shipping connections exist between the occupied spatial range and the range that may be invaded, the potential natural and human transport patterns need to be considered. NIS can be transported from the Ponto-Caspian region to the Great Lakes via an intermediate step in Western Europe. In addition to direct invasion pathways from the Ponto-Caspian region to Western Europe, MacIsaac et al. (2001) proposed four indirect pathways along the major rivers: the Danube and Rhine River pathway; (2) the Dnieper, Pripiat, Nemuna, and Vistula River pathway; (3) the Volga River system pathway; and (4) the Don and Volga River pathway. Many of these connections are completed through man-made canals and waterways which have allowed considerable exchange of species between water bodies (Reid and Orlova, 2002). To fully understand past indirect linkages (which, in turn, might help predict future indirect linkages) it would be necessary to have complete shipping data both from the Great Lakes to the intermediate port and from the intermediate port to the Ponto-Caspian region.

The natural construction of the Great Lakes, whereby water flows and boat traffic moves from one lake into another, facilitates natural and human-induced dispersal within and between the lakes (Duggan et al., 2003). These dispersal patterns are likely to hasten the spread of a NIS once it has entered the Great Lakes but are unlikely to add new species.

2.1.2. Ballast Water and NIS

Cargo vessels frequently take on ballast water to maintain stability when traveling from port to port and especially when crossing an open sea. Some or all of the ballast water is later released when cargo is loaded at various ports and, with regards to this study, those Great Lakes ports shown in Figure 1.

Ballast water is the largest source of NIS to the Great Lakes as shown in Figure 2. Additional sources of NIS to the Great Lakes include fish stocking programs, private aquaculture, the bait industry, the aquarium and ornamental pond industry, live fish food markets, recreational boating, and canals and diversions (Kerr et al., 2005).

Figure 1. The five Great Lakes, some of the Great Lakes ports, and surrounding region

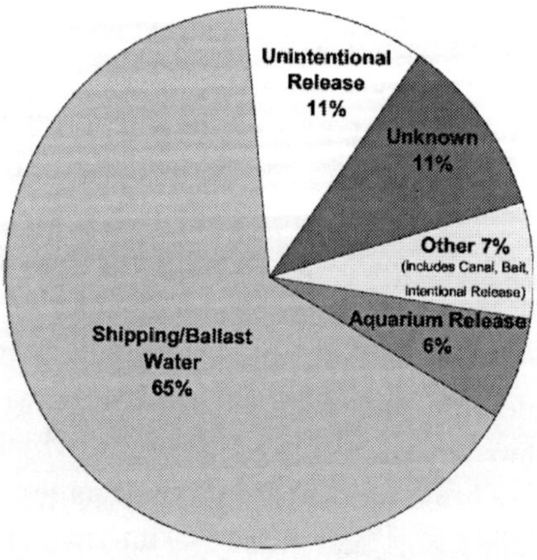

Figure 2. Sources of Great Lakes species invasions from 1960–2006 (based on data provided in Appendix A, which are derived from NOAA, 2007a).

While ballast water discharge (BWD) is the most prevalent pathway, an increase in BWD does not directly translate to more species invasions. Most discharges of ballast water in the Great Lakes occur in Lake Superior (Colautti et al., 2003), yet Lake Superior has less invasive species than any of the other Great Lakes (Grigorovich et al., 2003b). The low NIS colonization rate in Lake Superior may be due to any of several factors including cooler temperatures, a high ratio of deeper waters, low food availability due to low productivity, and low calcium concentrations (Grigorovich et al., 2003b).

2.1.3. Measures to Control the Release of Ballast Water Containing NIS

In response to NIS invasions stemming from ballast water releases in the Great Lakes, voluntary ballast water exchange (BWE) guidelines were implemented by Canada in 1989 and made mandatory in 2006. Mandatory BWE regulations were instituted by the USCG in 1993. These regulations require vessels carrying ballast water and entering the U.S. Great Lakes from outside the U.S. Exclusive Economic Zone (usually 200 miles away from the United States) to comply with one of the following three options:

(1) Vessels may exchange ballast water in open-ocean waters more than 200 nautical miles from any shore, and in waters more than 2,000 m deep, before entering the Snell Lock, at Massena, New York, provided that salinity of the ballast water is at least 30 parts per thousand (ppt).
(2) Vessels may retain their ballast water on board (vessels in this status are referred to as Ballast-on-Board, or BOB vessels).
(3) Vessels may use an alternative environmentally sound method of ballast water management that has been submitted to, and approved by, the Commandant of the USCG or an authorized representative before the vessel's voyage (33 CFR 151.1510).

Compliance with these ballast regulations has been high. From July 1999 to June 2001, 93% of regulated ships reported performing the necessary level of active BWE before arriving in Massena. The remaining 7% of ships were forced by the USCG to perform some sort of alternative action, such as decontamination, prior to being allowed to enter the Great Lakes (USCG, 2001). USCG reported high rates of BWE compliance (89 +/- 10%) for the period 1992–2004 (Ruiz and Reid, 2007).

Ballast water exchange at sea works first by the dilution effect. Assuming a homogenous distribution of flora and fauna in the ballast tank, 95–99% of the fresh water (and organisms) would be replaced by seawater (NRC, 2008). Second, BWE can be effective since most remaining freshwater organisms in the ballast tank are killed by the resulting high salinity levels.

Despite these ballast water regulations, at least 13 new NIS are believed to have entered the Great Lakes from ballast water since 1993 (Appendix A; IAGLR, 2002; Holeck et al., 2004; NOAA, 2007a). It is possible that BWE has been effective and that all the species found after 1993 were introduced before 1993; it just took many years to detect and report them (Costello et al., 2007a). Others have noted that organisms can survive BWE, and that BWE practices have not been completely effective in terminating the flow of NIS into the Great Lakes (e.g., Grigorovich et al., 2003a, b; Holeck et al., 2004; Drake et al., 2005; Ricciardi, 2006). Recently, more stringent regulations have been implemented (e.g., 73 FR

37, p.9950), which should reduce the flow of NIS into the Great Lakes from commercial shipping.

2.1.4. NOBOB Vessels and Species Invasions

Vessels fully loaded with cargo generally carry no ballast water on board. Vessels with no ballast-on board, commonly called NOBOB vessels, entering USGLP were not required to flush their ballast tanks or use an alternative treatment method until 2006. It is possible that invasions may have occurred from NOBOB vessels arriving in the Great Lakes (MacIsaac et al., 2002; Johengen et al., 2005). The almost completely empty ballast water tanks in NOBOB vessels often still contain residual sediment and water from previous ballasting operations. These residuals cannot be pumped from the ballast water tanks since the pump-out ports cannot be closer than several inches from the bottom of the tank. Residual material in ballast water tanks of NOBOB vessels can contain thousands of live organisms, their resting eggs and cysts, and microorganisms, including human pathogens, all of which may be discharged into Great Lakes waters (Johengen et al., 2005). When a NOBOB vessel off-loads cargo at a Great Lakes port of call it often takes on Great Lakes water into its ballast tanks to reestablish ballast. This pumped in freshwater mixes with the residual material in the ballast tanks, thereby increasing the viability of organisms. When such a NOBOB vessel then moves to another port to take on cargo, it may then discharge some or all of the recently acquired ballast water, along with the NIS from the earlier residual material.

NOBOB vessels currently account for about 90% of all inbound traffic to the Great Lakes (MacIsaac et al., 2002). Due to the number of potential invasive species in the residual material in NOBOB vessels and due to the large relative proportion of NOBOB vessels entering the Great Lakes, NOBOB vessels could pose a significant invasion risk. MacIsaac et al. (2002) found that for bacteria, copepods, cladocerans, and rotifers, NOBOB vessels may be exerting 10 to 100 times as much propagule pressure as vessels with ballast on-board complying with the regulations.

2.1.5. Other Options for Controlling Species Invasions from Ballast Water

As a result of the threat from NOBOB vessels, Canada developed mandatory regulations in 2006, requiring that transoceanic NOBOB vessels arriving in Canada undergo ballast flushing to eliminate fresh or brackish water residuals in their ballast tanks. Coastal vessels entering Canadian ports must comply with fairly similar requirements, only the BWE or ballast flushing must occur in an area only 50 nautical miles from shore (GLBWWG, 2008). Since August 2005, NOBOB vessels entering U.S. waters have been strongly encouraged, but not required, to conduct saltwater flushing before entering the Great Lakes (71 FR 18, pages 4,605–4,606). The St. Lawrence Seaway Development Corporation published regulations, which became effective at the start of the 2008 navigation season, requiring all NOBOB vessels that have operated outside the exclusive economic zone (usually 200 miles from the United States) to conduct saltwater flushing of their ballast tanks before transiting the St. Lawrence Seaway, regardless of whether their destination is a U.S. or Canadian port (73 FR 37, p. 9,950).

It is not yet possible to measure the effectiveness of recent regulations or guidelines because there is a time lag between when a species is transported, colonizes, and reproduces to a large enough population, to be detected and reported. However, the National Research

Council recommends that a binational science-based surveillance program be established to monitor for aquatic invasive N IS (N RC, 2008). The recommended program should involve dedicated lake teams, as well as academic researchers, resource managers, and local citizens groups, and it should leverage existing monitoring activities whenever possible.

NOAA is testing the effectiveness of BWE along with other various methods to treat ballast water using mechanical (e.g., filtration and separation), physical (e.g., sterilization by ultraviolet light, ozone, heat, electric current, or ultrasound), and chemical (e.g., chlorine dioxide) methods (NOAA, 2007b). The State of Michigan has established its own ballast water legislation, and other Great Lakes states are considering similar regulations (NRC, 2008). In 2007, the Michigan Department of Environmental Quality started prohibiting ballast water releases from oceangoing vessels into Michigan waters until a permit was issued by the state. Permits require one of four approved treatments, either sodium hypochlorite, chlorine dioxide, ultraviolet light radiation treatment preceded by suspended solids removal, or de-oxygenation (MD EQ, 2008). Because Michigan is currently an import state, there have been no permit applications to discharge ballast water into Michigan ports since Michigan's law was implemented in 2007 (telephone conversation on August 8, 2008 between Barry Burns, Michigan Department of Environmental Quality, and Vic Serveiss, U.S. EPA, NCEA). Therefore, oceangoing vessels visiting Michigan ports have not needed to install the Michigan approved ballast water treatment methods.

2.2. Conceptual Framework

Owing to the fact that invasive species are a major threat to ecosystems, there is a need to develop predictive tools and to demonstrate their use to natural resource managers as they consider ways to manage the problem. The approach used in this assessment, as shown in Figure 3, is based on Williamson‘s (1996) conceptual framework regarding biological invasions. Specifically, Williamson‘s thesis states that successful invasions are best predicted by knowing the propagule pressure (i.e., the number of larvae/individuals entering a new area) and matching the invaded habitat with the habitat in the invader‘s historical range.

2.2.1. Propagule Pressure

Propagule pressure is a composite measure of the number of individuals of a species released into a region to which they are not native. It incorporates estimates of the absolute number of individuals involved in any one release event (propagule size) and the number of discrete release events (propagule number). The probability of establishment of an introduced species increases as propagule pressure increases (Menges, 1998, 2000; Simberlof and von Holle 1999; Kolar and Lodge, 2001). In considering the sources of propagules to the Great Lakes, ballast water becomes the primary concern as shown in Figure 3.

The condition and life stage (resilient resting stages compared with sensitive juvenile stages) of propagules will also strongly affect the probability of establishment (Smith et al., 1999; Hayes and Hewitt, 2000; Wonham et al., 2001). Thus, management actions that reduce the number of released individuals, the number of introduction events, and the health of individuals released are likely to reduce the risk of invasion. Unfortunately, detailed quantification of these factors is limited and thus surrogate measures become necessary to estimate propagule pressure. As shown in Figure 3, the conceptual approach for this study uses ballast water discharge data as a surrogate measure for propagule pressure. Current scientific understanding of invasion biology suggests strongly that consideration of propagule pressure should be a major component of an assessment.

2.2.2. Habitat Suitability

2.2.2.1. Species Distribution Modeling

Assessing the degree to which a new environment is similar to the donor environment is a reasonable starting point to try to answer the question "Is a species likely to survive in this environment if it were introduced here?" Good computer-based tools are available that provide a first-cut broad geographical answer to the question. Standard methods for modeling suitable habitat include traditional multivariate statistical methods (e.g., discriminate analysis, multiple regression, logistic regression), often coupled with geographic information systems (GIS) (e.g., Ramcharan et al., 1992; Buchan and Padilla, 2000). More recent methods that are tailor-made for identifying potential ranges include CLIMEX (Sutherst et al., 1999), Genetic Algorithm for Rule-Set Production (GARP) (Peterson and Vieglais, 2001; Drake and Bossenbroek, 2004) all of which are embodied in user-friendly and readily available software. Predicting suitable habitat is also possible for aquatic environments (Drake and Bossenbroek, 2004; Marchetti et al., 2004a), but currently less tractable than for terrestrial habitats because: (1) fewer aquatic physico-chemical data are available in appropriate electronic formats, and fewer distribution data have been collected for aquatic species; (2) terrestrial climatic data are often poor predictors of the aquatic environment; and (3) strongly predictive environmental variables for establishment are unknown for many aquatic species (Carlton et al., 1995). All of these predictive models have at least two intrinsic limitations. First, environment matching assumes that no evolution will occur in the nonindigenous species with respect to habitat requirements (Cox, 2004; Sakai et al., 2001). Second, biotic interactions in a new environment may limit or facilitate establishment independent of any climatic match (Torchin and Mitchell, 2004).

2.2.2.2. Genetic Algorithms for Rule-Set Production (GARP)

GARP develops predictions of the potential geographic extent of an invasion by first modeling relationships between known occurrences of a species and the corresponding abiotic environmental variables, and then projecting the modeled species-environment relationships to a region of interest. GARP modeling requires two types of inputs: (1) spatial data describing the location of species based on occurrence data and (2) digital data layers describing environmental conditions at locations coinciding with the species occurrence data. GA RP develops outcomes consisting of a set of conditional rules in the form of if-then' statements that describe the ecological conditions of the species in its studied habitat (Stockwell and Peters, 1999). Habitats are matched by searching iteratively for nonrandom correlations between a species' known location and a variety of environmental parameters.

The GARP method is considered to be based on models of genetic evolution (Holland, 1975) because GARP models are built by an iterative process of rule selection, evaluation, testing, and incorporation or rejection of the rules produced (Peterson et al., 1999). With each iteration, rules are modified by selection, crossover, and mutation—resembling the genetic process. In the first phase, GARP selects a random population, based on a combination of initial prediction rules, which might represent suitable solutions for the problem. The fitness to the characteristics of the population is then evaluated for each pixel in the search space. If the performance of the rule is adequate as determined by the rule's significance measure, the rule is retained for further runs of the algorithm, until an end condition—consisting of a convergence limit and maximum number iterations—is satisfied (Stockwell and Peters, 1999). One of the main advantages of GARP is its ability to apply different types of rules at once to explain complex nonlinear relationships between the species occurrence and predictive variables. This implies that the algorithm can _learn' through each iteration and apply the type of rule that describes best the relationship among the variables for any particular portion of the search space (i.e., all possible combinations of variables) (Stockwell and Peters, 1999).

2.2.2.3. Modification of GARP for Aquatic Systems

GARP has been used to predict a variety of species distributions including birds in Mexico (Feria and Peterson, 2002; Stockwell and Peterson, 2002; Anderson et al., 2003) and North America (Peterson and Cohoon, 1999); rodents in South America (Anderson et al., 2002 and Anderson et al., 2003); and invasive vector disease insects in South America (Peterson et al., 2002). This may be the first time that remote sensing data were used in conjunction with GARP to predict invasive freshwater aquatic species.

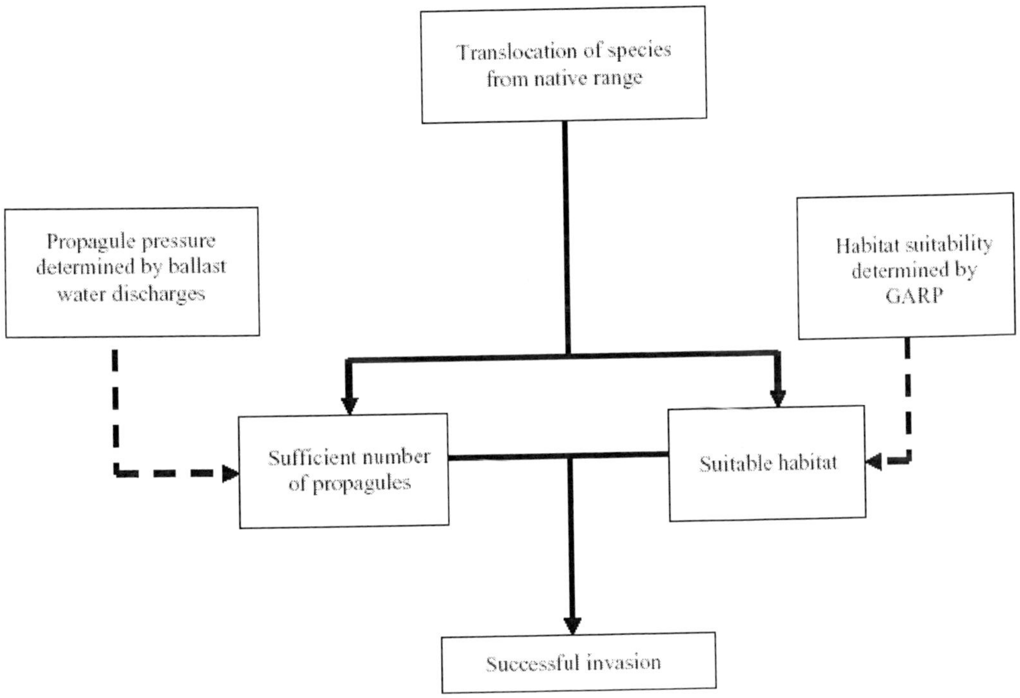

Figure 3. Conceptual framework for predicting future introductions of nonindigenous species into the Great Lakes of the United States

3. METHODS

For a nonindigenous species (NIS) to become established in the Great Lakes, the species must (1) move or be transported from its existing spatial range to the Great Lakes and (2) be able to colonize, become established, and spread in the new environment (Williamson, 1996; Theoharides and Dukes, 2007). Others have also combined these two analyses to predict NIS spread, though different names were used to characterize their respective efforts. Leung and Mandrak (2007) combined invasability and propagule pressure, to make predictions about zebra mussel spread. Herborg et al. (2007) combined introduction effort and environmental niche models to predict the potential spread of the Chinese mitten crab (*Eriocheir sinensis*) in North America. To address both requirements for successful invasion, we used information on ballast water discharges as a surrogate for propagulc pressure and the Genetic Algorithm for Rule-Set Production (GARP) model to determine the suitability of habitat by matching the invaded habitat in the Great Lakes with the species native habitat.

3.1. Habitat Suitability Using the GARP Model

Habitat suitability was modeled using a species distribution model to compare the environmental conditions associated with the distribution of invasive species in their home range with the conditions found in the Great Lakes. GARP was selected because it is a well established model, is one of the few models that accepts presence-only distribution data (e.g., locations where the species has been observed without corresponding information on where the species has *not* been observed), and incorporates multiple statistical approaches into a single framework.

3.1.1. Selection of Modeled Species

The first step to using GARP is to select the species to be evaluated. Species of interest (i.e., those thought to be potential invaders of the Great Lakes system) were identified based upon a review of the literature and best professional judgment. We searched for species' scientific names and the keywords "invasive" and "Great Lakes" in publications after 1990 using Web of Science and international databases, such as Fishbase (Froese and Pauly, 2007) and Global Invasive Species (IUCN, 2006). We augmented this general search strategy to include authors who have studied Great Lakes invasive species to find potentially relevant papers that did not specifically include the terms "invasive" or "Great Lakes" in the article's title, abstract, or keywords. Other sources include the U.S. Geological Survey, the States' Department of Natural Resources (for states adjacent to the Great Lakes), the Canadian Wildlife Federation, and the Great Lakes Panel on Aquatic Species. We initially identified 156 species of concern based on a review of the literature (see Appendix B).

Of the 156 species identified, using best professional judgment it was determined that 58 of these species pose the most risk for their potential to invade the Great Lakes and reach population levels that could cause ecological impact (see shaded entries in Appendix B). Twenty-eight of the 58 species identified are already in the Great Lakes. The remaining 30 species, not yet reported in the Great Lakes, were evaluated to see if sufficient data was available to run the GARP model. GARP requires at least 30 spatially unique occurrence points (i.e., latitude-longitude coordinates of locations where the species has been reported) to develop robust predictions (Stockwell and Peterson, 2002). For a variety of reasons, only 9 of the 30 species had sufficient data to be modeled. Of these nine, five species have not yet been detected in the Great Lakes, and the other four have been reported only infrequently. At the request of EPA's Great Lakes National Program Office, we modeled five additional species already found in the Great Lakes. Two of the five species, the zebra mussel and round goby, are currently widespread throughout the Great Lakes. The three other species, ruffe (*Gymnocephalus cernuus*), quagga mussel, and New Zealand mud snail (*Potamopyrgus antipodarum*), have been reported as established in the Great Lakes but are not yet widespread (USGS, 2007). Thus, a total of 14 NIS species were evaluated using the GARP model for the availability of suitable habitat in the Great Lakes (Table 1).

3.1.2. Model Inputs and Environmental Data Layers

3.1.2.1. Environmental Data Layers

Six specific parameters were used to define environmental variables suitable to develop data layers for GARP: mean, maximum, and minimum monthly surface water temperature; chlorophyll *a* concentration; the diffuse attenuation coefficient; and normalized water-leaving radiance (Table 2). These six parameters were chosen because they represent important environmental variables that tend to control the distribution of species. Three of the parameters are measures of temperature that affects species distribution worldwide. The other three are related to the productivity of aquatic systems. Chlorophyll *a* is an indicator of biological productivity. Water clarity, as measured by diffuse attenuation coefficient and the water-leaving radiance, is an indicator of the trophic state of the system. Water clarity also influences the depth of the photic zone and the ability of primary producers to acquire sunlight and flourish. Although some of these six data layers may be covariant, GARP is considered to be relatively robust to collinearity (Kluza et al., 2007). For the species modeled in this chapter, no literature reliably supports the a priori weighting of any of the selected environmental variables as more important than any other variable.

3.1.2.2. Environmental Data Sources

Water temperature was derived from the satellite-based Advanced Very High Resolution Radiometer (AVHRR) Oceans Pathfinder Sea Surface Temperature Data set and is accurate to within 0.5°C (http://podaac-www .j pl. nasa.gov/sst/). We used temperature data from 1985 through 2001. We used the MMT data to calculate three data layers for use by the GA RP models: Maximum MMT, Mean MMT, and Minimum MMT. Maximum MMT represents the highest value of each of the 12 sets of monthly averages of data. To calculate the mean MMT, we assigned each pixel the average of the 12 monthly averages. The minimum MMT represents the lowest value of each of the 12 monthly averages. Chlorophyll *a* concentrations were obtained from the Moderate Resolution Imaging Spectroradiometer (MODIS) sensor. The Diffuse Attenuation Coefficient (K490) relates to the presence of light-scattering organic and inorganic particles in the water column and is inversely related to water clarity (http://oceancolor.gsfc.nasa.gov/PRODUCTS/k490.html). The normalized water-leaving radiance is the radiance of reflected light at 551 nm. Since water absorbs very little light at

551 nm, increases are due to light reflection out of the water, which are usually caused by nonabsorbing particles such as suspended sediments.

Table 1. Fourteen Species Modeled Using GARP and the Source of Occurrence Data

Species common name and year reported	Description	Useful occurrence data records	Data source
Species already widespread in the Great Lakes			
Gymnocephalus cernuus, ruffe—1986	fish	229	GBIF[a] USGS[b]
Dreissena polymorpha, zebra mussel—1988	mollusk	268	GBIF, USGS
Dreissena bugensis, quagga mussel—1989	mollusk	83	USGS
Neogobius melanostromus, round goby—1 990	fish	145	GBIF, USGS
Potamopyrgus antipodarum, New Zealand mud snail—1991	mollusk	867	GBIF, USGS
Species reported in the Great Lakes but either not extensive or lacking spatial data			
Cercopagis pengoi, fishhook waterflea—1 998	crustacean	152	GIS[c]
Scardinius erythrophthalmus, rudd—1989	fish	57	GBIF, CIMS[d]
Proterorhinus marmoratus, tubenose goby—1990	fish	171	CIMS, BSRDB[e]
Alosa aestivalis, blueback herring—1995	fish	408	GBIF
Species not yet reported in the Great Lakes			
Corophium curvispinum, N/A	amphipod	65	GBIF, CIMS
Neogobius fluviatilis, monkey goby	fish	50	CIMS
Pomatoschistus minutus, sand goby	fish	102	GBIF, BSRDB
Rutilus rutilus, roach	fish	117	GBIF, CIMS
Tinca tinca, tench	fish	50	CIMS

[a] G lobal Biodiversity Information Facility, 2007 (http://www.gbif.org/).
[b] USGS Nonindigenous Aquatic Species Database, 2007 (http://www.usgs.gov/pubprod/maps.html).
[c] Regional Biological Invasions Center. INVADER, 2007 (http://www.zin.ru/proj ects/i nvasions/gaas/invader/invader.htm).
[d] Caspian Interactive Map Service, 2007 (http://ipieca.unep-wcmc.org/imaps/ipieca/caspian/viewer.htm).
[e] Black Sea Environment Programme Red Data Book, 2007 `(http://www.grid.unep.ch/bsein/redbook/index.htm).

3.1.3. Use of GARP model

A stand-alone version of Desktop GARP (version 1.1.6) (Scachetti-Pereira, 2002) was used to model the distributions of the first nine N IS; Desktop GARP (version 1.1.6) within Open Modeler (version 1.0.5) was used for five of the species. There is no functional difference between the two desktop versions used. GARP relies on species occurrence or presence data and synthetic species absence data, termed pseudo-absence data. The use of pseudo-absence data is an intrinsic and accepted part of GARP modeling. To develop pseudo-absence data, investigators must identify a region surrounding the occupied range of the target species to which the species could easily spread. It is assumed that the reason the species is absent from the surrounding region is because environmental conditions are different and outside the species tolerance limits. The established range includes the species' occupied range and a surrounding unoccupied range. The pseudo-absence data are selected from a subset of locations within the investigator-defined study area that are not currently occupied by the species. The pseudo- absence points represent the locations presumed to be unsuitable for the target species and provide a contrast against which occurrence models can be developed. A new, random selection of pseudo-absence data was made for each model iteration.

GARP divides the occurrence data into training and test data sets. Test data sets are reserved to test predictive performance of models that are developed using the training data sets. GA RP then uses the training data and one of the individual algorithms to develop a model, and the model is tested and improved until the best solution is found. For each of the 14 modeled species, GARP randomly assigned the data into 50/50 splits of training and test data sets. We produced 1,000 model runs from the training data sets (i.e., 14,000 total GARP runs) that are all slightly different and vary in predictive ability. Each individual GARP model run produces a map of '0s and 1s,' representing predicted absence and presence, respectively. The area of predicted presence for each run is simply the proportion of pixels that have a predicted value of 1. We used a procedure described by Anderson et al. (2003) to select the best subset of runs generated for each species to develop a final composite range map. The model is run for each species and compared to see which runs best predicted the known occurrence locations and omitted the fewest of the known occurrence locations of the modeled species. For each species, we retained the models with the lowest omission error rates when compared to the results of the test data sets.

Table 2. Environmental Variables Used to Predict Locations That Would Provide Suitable Habitat for the 14 Modeled Species in the Great Lakes. The Spatial Resolution of Each of These Six Data Layers is ~21 km^2

Variable	Units	Source	Collection period
Mean monthly temperature	°C	AVHRR[a]	1985–2002
Maximum mean monthly temperature	°C	AVHRR	1985–2002
Minimum mean monthly temperature	°C	AVHRR	1985–2002
Chlorophyll *a* concentration	mg/m^3	MOD IS[b]	2001–2005
Diffuse attenuation coefficient (K490)	m^{-1}	MODIS	2001–2005
Normalized water-leaving radiance (nLW551)	mW/(cm^2 µm sr)	MODIS	2001–2005

[a] Advanced Very High Resolution Radiometer.

[b] Moderate Resolution Imaging Spectroradiometer.

Next, the median area of predicted presence for each species (simply the number of pixels that have a predicted value of 1) was determined from the models with the lowest error rates. The subset of models within 50% of this median value was then selected to form the final prediction of habitat suitability. Every pixel on the model output represents either a predicted presence or a predicted absence. The final composite prediction maps reflect the sum of the results from the models; that is, each pixel was assigned a value ranging from zero (i.e., no models predict presence) to 100 (i.e., all models predict presence). This value represents the relative environmental suitability of each location.

The final habitat suitability maps reveal the frequency with which a pixel is predicted to provide suitable habitable and depict the repeatability of that prediction with different models developed from different, randomly divided, training data sets. The higher the value of a pixel, the more likely the modeled species is expected to find suitable habitat at that location. Another interpretation is that pixels that have higher values represent higher quality habitat for the modeled species because these locations are predicted to be suitable by more models. Those pixels that are predicted to provide highly suitable habitat within the Great Lakes are characteristic of environments similar to those known to be occupied by the modeled species in their natural range.

3.1.4. Assumptions and Limitations

The use of species distribution models to predict the spread of NIS required three assumptions: (1) the available distribution data describe the full range of environmental conditions that the modeled species can tolerate; (2) the environmental variables selected to model potential spread govern the current and future geographic ranges of the NIS under study; and (3) biotic factors do not influence species distributions, unless such biotic factors can be included explicitly as environmental data layers. Failure to meet these three assumptions can limit the ability of GARP to predict invasive spread and can result in two types of errors, which arise from two broad sources: (1) limitations inherent in the data or the model themselves (these are often termed *data errors*), and (2) ecological processes relevant to the distribution of the species that are not included in the model (often called *biological errors*).

3.1.4.1. Data Errors

A major source of error in GARP modeling is the lack of complete occurrence data for the modeled species. GARP requires occurrence points that are both representative of the full range of environmental conditions associated with presence of the species and of the area inhabited by the species. In reality, the actual occurrence points reflect bias in both sampling and reporting efforts, which is influenced by resources, accessibility, transportation corridors, and visitation frequency. Further, the occurrence data extracted from online databases were collected *ad hoc*, and not for the purposes of constructing distribution models. If occurrence data do not describe the full environmental tolerances of the species, predictions will underestimate areas where a NIS could survive and establish itself in a particular region of the Great Lakes. For example, occurrence data for the monkey goby and the tench were available only for their distributions within the Iranian portion of the Caspian Sea. This may be part of the reason these two species have the smallest predicted distributions of any modeled species within the Great Lakes, perhaps suggesting under-prediction. Therefore, the predicted habitat suitability might not include all environments which the NIS could invade.

Some occurrence data were discarded because they did not fall in waters defined by the 21 km^2 spatial resolution available for the environmental data layers. These discarded occurrence points are more than likely in lakes and rivers smaller than 21 km^2 that could not be resolved by the satellite sensors used in this study. GARP requires that species absence data be developed by accurately selecting the region for which the species is absent because environmental conditions are different and outside the species tolerance limits. Determining the extent of the GARP prediction region assumes that these pseudo-absence points really are uninhabitable, and not, for example, simply suitable environments to which the target species has not yet dispersed.

Model errors can also result from modeling habitat suitability with a limited set of environmental variables. While we know that each of the six selected variables has a strong influence on species distributions, other abiotic factors known to influence species distribution are not captured by the variables that we used. For example, salinity impacts the survival of many aquatic organisms (Bailey et al., 2005), but salinity is not included in the GARP analysis. We were unable to locate a global database with spatial salinity data at the same scale of resolution as the six variables included in this study. Including salinity at a coarser resolution would have introduced coarse range boundaries where salinity was the limiting factor. Bathymetry data were available, but not used, because species occurrence data did not include the depth at which the species was found. As many aquatic species may only survive in waters to a certain depth, the model would show some deeper waters as suitable habitat and may contradict what is known about the depth limitations of a particular species. Nutrient levels may also be biologically important to some NIS. Calcium concentrations, for example, are likely a limiting factor for zebra and quagga mussel distributions (Cohen, 2007), but global spatial databases of calcium concentrations are not available. Failure to include such key factors can lead to over-predictions (predicting that the species can survive in an area where the habitat is actually unsuitable due to the environmental variable not included in the model). Detailed knowledge of the target species is required to reliably determine if one of more key factors have been excluded from the models. Because the consequences of under-prediction (failing to identify a place where NIS can establish) are much greater than those resulting from over-prediction, our approach is conservative and errs on the side of over-prediction.

3.1.4.2. Biological Errors

Even if the environmental variables could accurately reflect the abiotic factors controlling species distributions, the predictions are developed without considering biotic factors such as competition, predation, and parasitism. Biotic factors also are important determinants of the distributions of species, but it is not ordinarily possible to obtain data on biotic factors for incorporation into GARP, and failure to consider such factors can lead to poor predictions (e.g., Fitzpatrick et al., 2007). In a new environment, a species may be freed of restrictions (e.g., a predator may not exist), encounter new challenges in a new environment (e.g., competition from a species with a similar niche), or evolve and adapt. Thus, it is difficult to predict whether the impacts of excluding biotic factors would inhibit establishment or expand the colonization range of an introduced species. The fire ant in the southeastern United States is an example of an introduced NIS that established itself beyond the predicted range of a species distribution model, perhaps due to biotic factors that encouraged the species successfulness (Fitzpatrick et al., 2007).

3.1.5. Testing the GARP Model Performance

Despite the limitations described above, species distribution models are currently one of the few techniques readily available to predict the potential for an invasive NIS to become established in an area of interest (Peterson, 2003). Therefore, species distribution models should be considered a key component of a multi-faceted NIS prevention and management plans (Mack, 1996; Peterson and Vieglais, 2001).

The GARP model outputs were validated by testing the ability to correctly predict independent data that were not used to develop the model. Specifically, we evaluated how well GARP performed by assessing how well the model predicted the known distributions of three NIS that are already widespread throughout the Great Lakes using distribution data collected outside of the Great Lakes. Thus, occurrence data for the zebra mussel, ruffe, and New Zealand mud snail within the Great Lakes were withheld from the GARP model runs and the model tested for its ability to correctly predict suitable habitat. The performance of the GARP model was assessed using area under the curve of the Receiver Operating Characteristic curve (Sing et al., 2005). Area under the curve is a threshold-independent evaluation of model performance that measures the ability of the model to differentiate between sites where a species is present from sites where it is considered absent. Area under the curve represents the probability that, when a predicted-present site and a predicted-absent site are drawn at random, the predicted- present site will have a higher predicted value. The effectiveness of the GARP modeling is based on the scale for determining model performance devised by Swets (1988). More details on the model validation approach we used are provided in Appendix C.

3.1.6. Determining GARP's Power to Predict

GARP and other species distribution models make predictions about the suitability of habitat for a particular species within a region of concern. These models are developed by comparing the environmental conditions in the region containing the species to those found in the region of concern, in this case the Great Lakes. As noted previously, predictions from GA RP and other species distribution models are valid only for the range of environmental conditions on which the model was developed. Reliable predictions cannot be made for any environment within the Great Lakes that are not similar to those found within the region containing the distribution of the study species. GARP does not have a method for determining when a reliable prediction cannot be made, and, instead, may report such environments as a predicted absence when they may indeed be habitable by the NIS. Reporting such areas as unsuitable habitat may be erroneous and could misdirect management attention away from these potentially susceptible areas.

We used a technique called "power-of-prediction analysis," devised expressly for this project, to distinguish between areas of predicted absence from areas for which a reliable prediction cannot be made (null prediction). Like GARP models for individual species, power of prediction analysis uses GARP to develop predictions. However, instead of developing a model of environments represented by the distribution of the study species, power of prediction analysis attempts to (1) model *all* environments within the region containing the distribution of the species and (2) compare these environments with those characterizing the Great Lakes.

To perform a power of prediction analysis, we identified a region encompassing the full range of environmental conditions to which the species is known to occur. For example, consider a hypothetical species in the Caspian Sea reported in regions with water temperatures between 15 and 20°C but not reported in regions with a water temperature from 10 to 15°C. GARP would then predict that places in the Great Lakes with water temperatures between 15 and 20°C provide suitable habitat and that all regions with temperatures less than 15°C are unsuitable. However, temperatures in the Great Lakes range from 6 to 20°C. This presents a problem regarding areas within the Great Lakes that range from 6 to 10°C, which is below any temperature found in the Caspian Sea. When considering temperature, in an isolated, univariate way, it is likely the species would not tolerate temperatures from 6 to 10°C if it could not tolerate 10 to 15°C. Yet, it is not correct to assume that all locations beyond a particular extreme, in this case less than 10°C, are unsuitable. The species experiences the environment in a multivariate manner and that could produce surprising and counterintuitive results. For example, a terrestrial species might be able to survive and reproduce in locations that were hotter, if they were also wetter.

Even though GARP cannot make a reliable prediction for such areas, GARP and many other species distribution models will report areas with temperatures outside of the range from 10 to 20°C as unsuitable for the species, when in reality the GARP model has no information to draw such a conclusion (Heikkinen et al., 2006). We used power of prediction analysis to denote the geographic extent of predictive power. We performed power of prediction analyses for 11 of the 14 modeled species. Power of prediction analysis was not performed for two of the invasive species already established in the Great Lakes—quagga mussel and round goby—due to the lack of occurrence data outside the Great Lakes. Also, no power of prediction analysis was needed for the blueback herring because GARP model runs predicted that the blueback herring can encompass essentially the entire area of the Great Lakes. Appendix D provides more details on how we applied the power of prediction analysis to this study.

3.2. Determining Propagule Pressure Using Ballast Water Discharge Data and Vessel Traffic Patterns

The probability that a NIS can become established increases with increased propagule pressure (Simberloff and Von Holle, 1999; Kolar and Lodge, 2001; Lockwood et al., 2005). Propagule pressure, as explained in the introduction, is the number of individuals (including larvae, seeds, and spores) released in a nonnative region over a specified period of time (Simberloff and Von Hol le, 1999). We used two sources of data as a surrogate for propagule pressure: Data from the U.S. Coast Guard's (USCG's) National Vessel Movement Center (NVMC) and the National Ballast Information Clearinghouse (NBIC). Ultimately, the NBIC data proved to be the most useful in predicting propagule pressure. The NB IC collects, analyzes, and interprets data on ballast water management practices of commercial ships that operate in the United States. NBIC was created by the USCG and the Smithsonian Environmental Research Center (NBIC, 2008) pursuant to the National Invasive Species Act of 1996 (16 USC 67 § 4712). NBIC‘ s data are electronic and are accessible on the Internet (http://invasions.si .ed u/nbic/).

3.2.1. Analysis of Ballast Water Discharge Data

The principal aim of the NBIC database is to quantify the amounts and origins of ballast water discharges in U.S. coastal systems and to determine the degree to which such water has undergone open-ocean exchange or alternative treatments designed to reduce the likelihood of ballast-mediated invasions by exotic species (NBIC, 2008). NBIC data come from national ballast water management reporting forms submitted to the USCG by vessels arriving to ports and places in the United States. The data includes port of arrival, date of arrival, and last port of call, along with the source of ballast water (either a specific port or a latitude/longitude coordinate at sea), date of ballast water intake, type of ballast water management, date discharged, and the volume discharged.

This database allowed us to locate the source of ballast water and to determine those Great Lakes ports receiving the most ballast water discharges with the most potential to transport NIS. NBIC data from for 2004–2007 for Illinois, Indiana, Michigan, Minnesota, New York, Ohio, Pennsylvania, and Wisconsin were downloaded and analyzed using a relational database (Microsoft® Office Access). The original data set contained records of 44,461 vessel arrivals and 121,031 ballast water discharges. By excluding records of vessels arriving in ports outside the Great Lakes system, the NBIC data set was reduced to 63,574 ballast water discharges.

Since NIS that were in the ballast tank before ballast water exchange at sea may survive the exchange and can later be released in the Great Lakes, we needed to determine the original source of ballast water. Discharges of ballast water that originated within the Great Lakes (which we defined as west of Quebec City, see Figure 1) was excluded along with discharges of ballast water that was derived from outside 200 nautical miles of any shore and deeper than 2,000 m. We analyzed the remaining 618 ballast water discharges because these waters have the most risk of transporting NIS. We identified the most common original source of ballast water and the U.S. Great Lakes ports (USGLP) receiving the most discharges.

As discussed previously, some vessels enter the St. Lawrence Seaway declaring to have no ballast on board (NOBOB) but as they traverse the Great Lakes, they take on ballast water which can mix with residual water or sediment in the ballast tanks. We combined NBIC with NVMC data since the later includes information on the last five ports of call. Starting with NB IC records, we matched the NB IC arrival port and arrival date to the corresponding data in the NVMC data set. For each of the vessel records with matching sets of data, we obtained the last five ports of call records from the NVMC database. If one of more of the last five ports of call were not in the Great Lakes, we considered the vessel to have entered the Great Lakes with no ballast on board but containing residual material (NOBOB-RM). We then calculated the number of ballast water discharges and the volume of ballast water discharged from each of these vessels.

3.2.2. Assumptions and Uncertainty

Although NBIC employs a rigorous quality assurance and quality control protocol, the accuracy and completeness of the self-reported data cannot be guaranteed (NBIC, 2008). The

NVMC data set also has limitations. Although a vessel may have stopped at a foreign port during one of its last five ports of call it does not necessarily mean that ballast water was taken on at that port. It is possible that any residual material in the ballast tank may be from within the Great Lakes and not the foreign port of call included in the last five ports of call records. This would lead to over-predicting the potential for NIS release. Similarly, it is also possible that we missed a source of residual material because it may have been picked-up sometime earlier than the last five ports of call. This would lead to an under-estimation of NIS releases. Finally, we assumed that the data from 2006–2007 used in this study are representative of discharge and shipping patterns over the past several years.

4. Results

We first present results comparing the Great Lakes with the rest of the world, especially the Ponto-Caspian region with respect to six environmental parameters used to model species distribution. We next identify those locations within the Great Lakes that would provide suitable habitat for each of the 14 modeled nonindigenous species (NIS). And third, we identify the ports within the U.S. Great Lakes that received the most ballast water discharges from the vessel traffic we analyzed, including the identification of ports around the globe that provided the original source of ballast water discharged at a U.S. Great Lakes port.

4.1. Comparison of the Great Lakes to the Ponto-Caspian Sea

The Ponto-Caspian region has been identified as a significant source of nonindigenous species entering the Great Lakes. The comparison of the Great Lakes to the Ponto-Caspian region, based on the six environmental parameters used in the Genetic Algorithm for Rule-Set Production (GARP) modeling, reveals that the regions are indeed quite similar. Figures 4–7 illustrate the environmental conditions for those parameters used in the habitat suitability modeling as shown in Table 2. Latitudinal differences and discernable patterns in deeper open ocean waters are clearly evident.

4.1.1. Temperature

Overall, the maximum monthly temperature (MMT) shows a strong latitudinal gradient (Figure 4). The Great Lakes, shown mostly in yellow, have a spatial mean of 9.9°C and range from 5.9°–1 3.7°C. Lake Superior is colder than the other four lakes and has a spatial mean of 7.9°C with a range of 5.9°–11 .1°C. The Caspian and Black Seas are somewhat warmer and less variable (in terms of maximum MMT) than the Great Lakes, with a spatial mean of 13.7°C and a range of 12.1°–16.1°C. The impact of climate change will likely cause the Great Lakes to reach MMT levels even more similar to the Ponto-Caspian Sea region (IPCC, 2007).

4.1.2. Chlorophyll a Concentrations

Due to upwelling, the western edges of continents display high concentrations of chlorophyll *a* (dark blue color in Figure 5), a surrogate measure of productivity. Colder arctic waters are also more productive. The global spatial mean concentration of chlorophyll *a* is

0.24 mg/m^3. The Great Lakes, shown mostly in blue in Figure 5, show a mean spatial productivity level of 1.7 mg/m^3, ranging from 0.19–62.3 mg/m^3. The Caspian and Black Seas are similar to the Great Lakes with a spatial mean of 2.2 mg/m^3 and range from 0.2–56.2 mg/m^3.

4.1.3. Diffuse Attenuation Coefficient (K490)

Overall, open ocean waters are generally clearer than waters with higher biological productivity. The more light that is scattered from the surface, the Diffuse Attenuation Coefficient (K490), the greater the amount of suspended solids, a measure of productivity. Greater K490 values imply more light attenuation and lower water clarity. The global mean K490 is 0.032/m, which translates to a photic zone depth of ~144 m. The K490 of the Great Lakes, shown mostly in blue in Figure 6, is much higher than the global average and has a spatial mean of 0.099/m (equivalent to a photic zone depth of ~47 m) and ranges from 0.037–0.741/m. The Caspian and Black Seas also are fairly turbid and similar to the Great Lakes with a spatial mean of 0.11/m (photic zone depths of ~42 m) and range from 0.05–0.72/m.

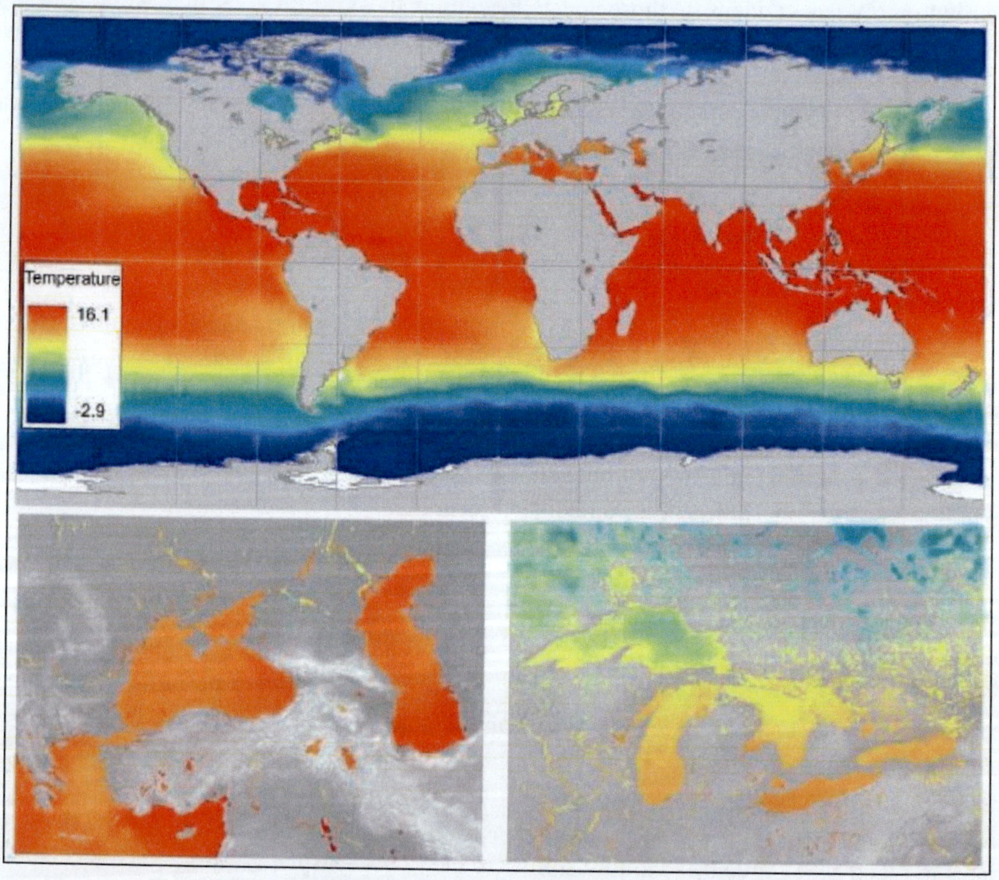

Figure 4. Maximum monthly mean temperature (MMT) (°C) as determined by AVHRR sensor (1985–2001). Warmest temperatures are indicated by red; cooler temperatures are indicated by shades of blue. Global view (top), Ponto-Caspian region (left), and Great Lakes (right)

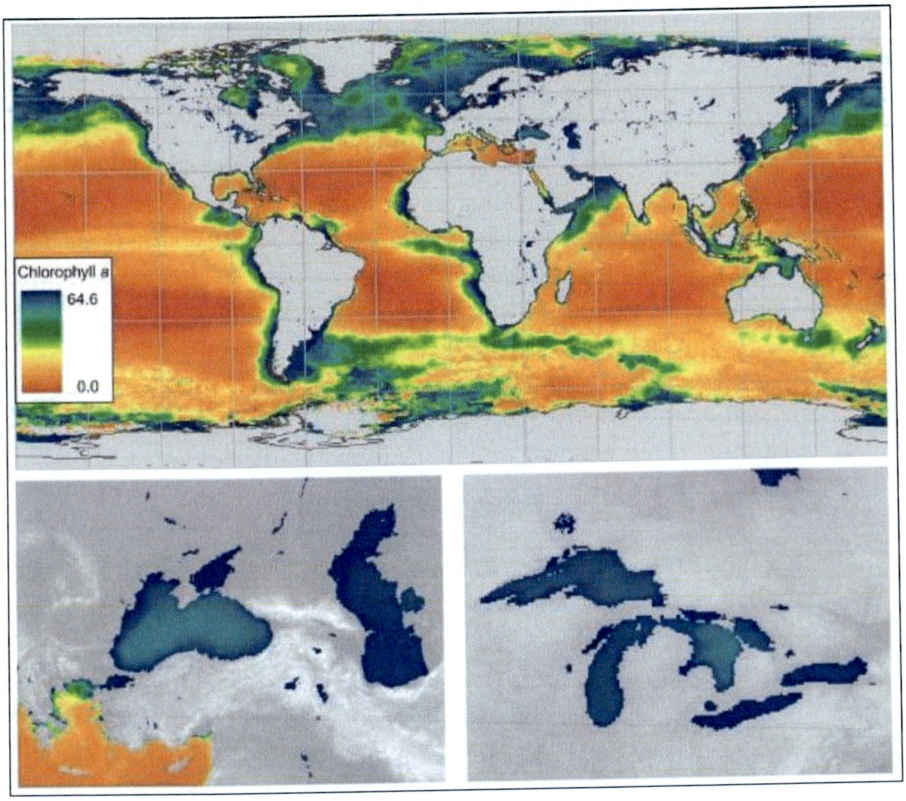

Figure 5. Average chlorophyll *a* concentration (mg/m^3) as determined by MODIS (2001–2005). High chlorophyll *a* concentrations are represented with blue and dark green. Brown and yellow indicate low concentrations of chlorophyll *a*. Global view (top), Ponto-Caspian region (left), and Great Lakes (right)

4.1.4. Normalized Water-Leaving Radiance

As with the Diffuse Attenuation Coefficient, this is a measure of the productivity of waters. As shown in Figure 7, and consistent with the previous figures, waters near the continents are generally more productive. The Great Lakes have a spatial mean of 1.5 mW/(cm^2 μm sr) and a range of water clarity from 0.0–5.1 mW/(cm^2 μm sr). Lake Superior, being much deeper (averaging 147 m and with a maximum depth of 406 m) than the other Great Lakes, has a spatial mean of 0.5 mW/(cm^{2}μm sr) indicating that Lake Superior is less productive. The Caspian and Black Seas have a spatial mean of 2.4 mW/(cm^2 μm sr), similar to the lower Great Lakes.

4.2. Habitat Suitability for Modeled Species

The results of using the GARP species distribution model reveals that the Great Lakes offers suitable habitat for all of the 14 modeled species in this study, with Lakes Erie and Ontario the most likely to be invaded. Five of the species modeled are already established in the Great Lakes and the remaining nine, selected from an original list of 156 species, are most likely to invade and become established in the Great Lakes. Figures 8 through 10 illustrate the suitability of habitat within the Great Lakes for the blueback herring (*Alosa aestivalis*), quagga mussel (*Dreissena bugensis*) and round goby (*Neogobius melanostromus*). Unfortunately, due to limits with occurrence data, a power of prediction analysis could not be performed for the quagga mussel and round goby. There was no reason to perform a power of prediction analysis for the blueback herring because it is predicted to find suitable habitat throughout the Great lakes. Figures 11 through 21 show the suitable habitat for the fishhook waterflea (*Cercopagis pengoi*), zebra mussel (*Dreissena polymorpha*), ruffe (Gymnocephalus *cernus*), monkey goby (*Neogobius fluviatilis*), New Zealand mud snail (*Potamopyrgus antipodarum*), tubenose goby (*Proterorhinus marmoratus*), rudd (*Scardinius erythrophthalmus*), an amphipod (*Corophorum curvispinum*), sand goby (*Potamoschistus minutus)*, roach (*Rutilus rutilus*), and tench (*Tinca tinca*). These figures also show areas where no reliable prediction could be made based on a power of prediction analysis.

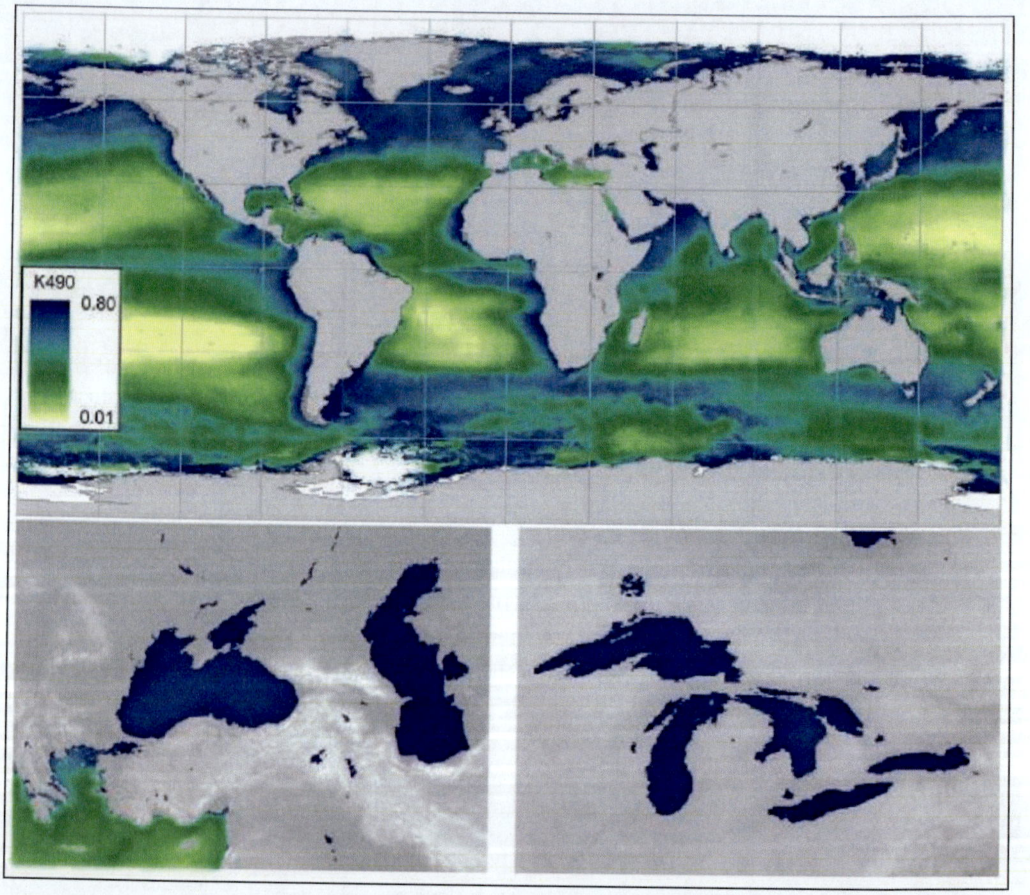

Figure 6. Average diffuse attenuation coefficient (m^{-1}) at 490 nm as determined by MODIS (2001–2005). Yellow and green colors indicate less light absorption, blues indicate greater attenuation of light. Global view (top), Ponto-Caspian region (left), and Great Lakes (right)

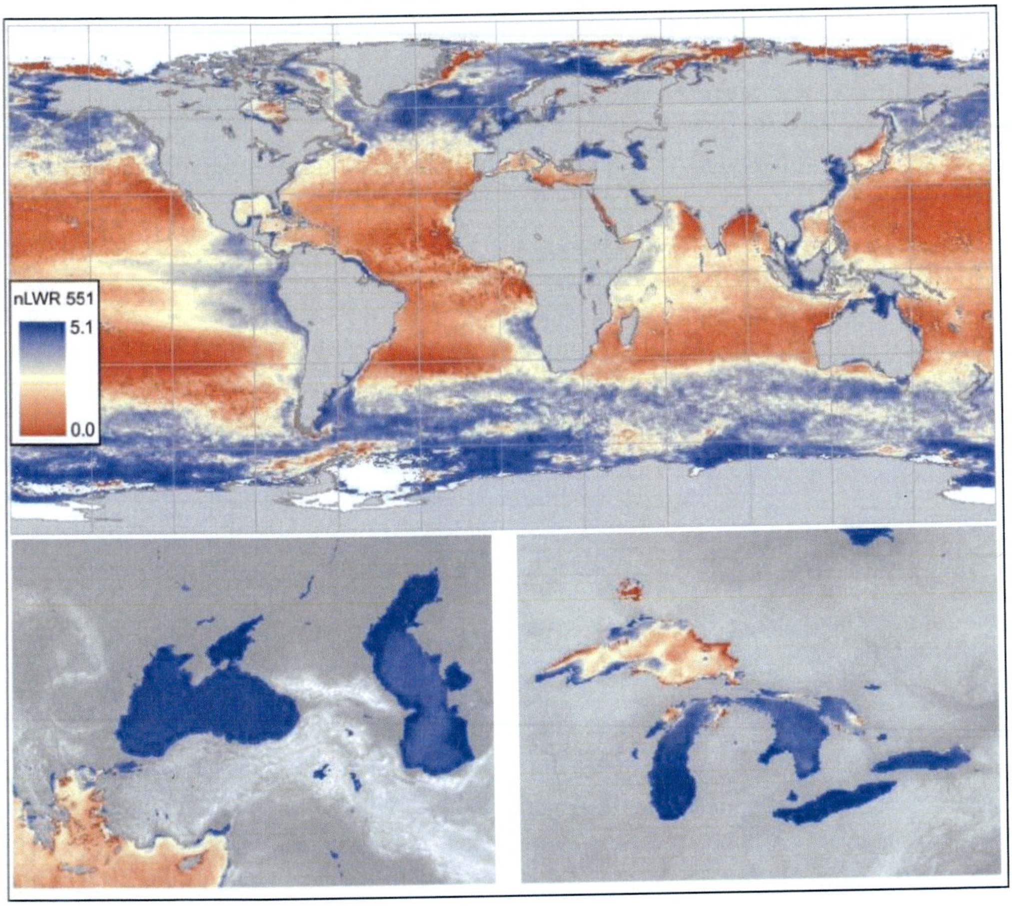

Figure 7. Average normalized water leaving radiance (mW/cm^2 μm sr) as determined by MODIS (2001–2005). Blues are higher values (i.e., higher concentrations of particles in the water which reflect more light), and reds and yellows indicate lower values (i.e., less light is emitted). Global view (top), Ponto-Caspian region (left), and Great Lakes (right)

The habitat suitability maps are based on the best 100 of 1,000 model runs for each species. The color scale (from blue to red) in the figures reflects the number of GARP model runs, from 0 to 100, that predict the modeled species would find suitable environmental conditions in the location being considered. Specifically, the red colored regions indicate where nearly all GARP model runs predicted that NIS would find suitable habitat. The blue-colored areas indicate where few or no GARP models runs predicted that NIS would find suitable habitat. The power of prediction analysis helps to distinguish between areas with a low likelihood of providing suitable habitat from areas where a prediction could not reliably be made. The gray regions denote areas where no reliable prediction can be made about the potential distribution of an invader. The color scale does not imply any measure of credibility or precision, but rather expresses commonality among predictions developed via a stochastic process of model generation.

To further validate the GARP model, results for the three species already reported in the Great Lakes (zebra mussel, ruffe, and the New Zealand mud snail) were compared with their current spatial distribution. This analysis indicates that GARP modeling is a good predictor of habitat suitability according to the scale for evaluating the performance of species distribution models devised by Swets (1988). The model performance scores are 0.79 for the zebra mussel and ruffe and 0.74 for the New Zealand mud snail. These scores, representing the area under the curve of predicted accuracy (Sing et al., 2005), suggest that the six environmental data layers that were selected as inputs for the GARP modeling are appropriate for predicting the locations that would provide suitable habitat. Appendix C provides more information on model validation.

4.2.1. Blueback Herring

If the blueback herring, a medium-sized fish, enters the Great Lakes it is very likely to find suitable habitat throughout the Great Lakes system, according to GARP (Figure 8). Only the deeper portions of Lake Superior and other isolated spots in other Lakes may not provide suitable habitat for this species. Without a power of prediction analysis, it is not possible to know if the blue colored areas reflect unsuitable habitat or areas where no prediction is possible. The blueback herring and alewife are of similar shape and general appearance, and distinguishing between them is difficult. Bluebacks tend to have a smaller eye than alewives, with the eye diameter usually smaller than the snout length. As their name implies, these fish often have dark blue backs. An anadromous fish, the blueback herring spends the greater part of its life in salt water and returns to fresh water to spawn. It usually spawns later in the spring than the alewife, when water temperatures are a bit warmer. During spawning, many eggs are deposited over the stream bottom where they stick to gravel, stones, logs, or other objects. A few surviving, spent fish move back to the sea after spawning. Young fish usually move to sea when about l month old and 1 1/2 to 2 inches long. Bluebacks feed on plankton, various small floating animals, small fish fry, and fish eggs. Although the Great Lakes are distant from marine waters, the blueback herring can spend its whole life and develop reproducing populations entirely in freshwater (VA Inland Fisheries, 2008). If blueback herring became established in the Great Lakes, they could impede recovery of depressed populations of indigenous fishes such as cisco and lake trout (Owens et al., 1998).

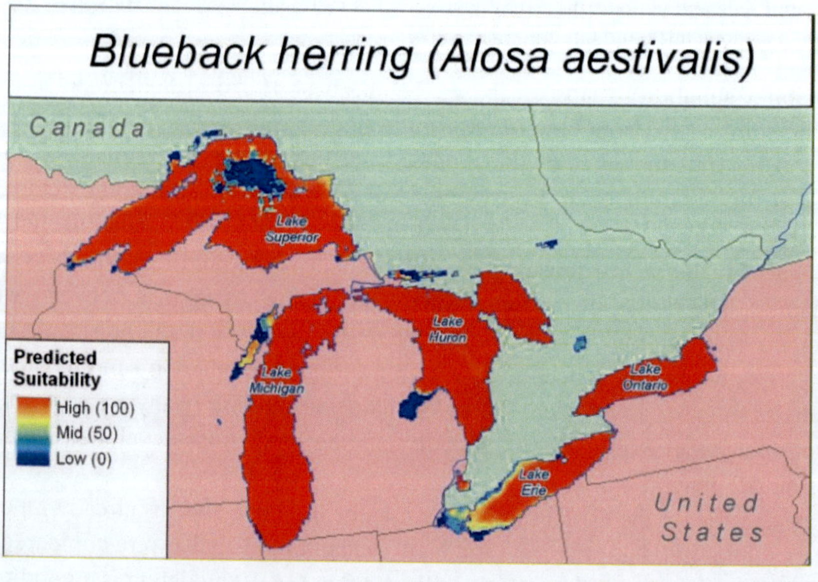

Figure 8. GARP-predicted habitat suitability of blueback herring (*Alosa aestivalis*) in the Great Lakes

4.2.2. Quagga Mussel

The quagga mussel, a mollusk, already occupies most shoreline areas in Lakes Erie and Ontario and southern Michigan (Figure 9). According to GARP modeling, the rest of Lake Erie and the southern shoreline zones of Lakes Michigan and Huron also are likely to provide suitable habitat for this species. Without a power of prediction analysis, predictions cannot reliably be made for the other regions. However, the species has already been reported in several of these locations, including the shorelines of Lake Michigan, Lake Superior, and Lake Huron. Quaggas are prodigious water filterers, removing substantial amounts of phytoplankton and suspended particulate from the water. As such, their impacts are similar to those of the zebra mussel. By removing the phytoplankton, quaggas in turn decrease the food source for zooplankton, therefore altering the food web. Impacts associated with the filtration of water include increases in water transparency, decreases in mean chlorophyll *a* concentrations, and accumulation of pseudofeces (Claxton et al., 1998). Quagga mussels prefer silty or sandy lake bottoms and can live in warm or cold water. MacIsaac (1994) correctly speculated that the quagga mussel was still expanding its nonindigenous range in the Great Lakes. It has spread to depths greater than it occupies in its native range (Mills et al., 1996) and is abundant to a depth of 150 m (Wisconsin DNR, 2008) and 174 m in Lake Ontario (Watkins et al., 2007). By 1999, the quagga mussel dominated southern Lake Ontario, where the zebra mussel was once dominant (Mills et al., 1999), and it continues to spread into regions previously occupied by the zebra mussel (Watkins et al., 2007). The ability to spread to areas that can be potentially occupied by the zebra mussel further supports the notion that spread and colonization may occur until the species reaches its depth limitation.

4.2.3. Round Goby

The GARP model predicts the round goby, a medium-sized, bottom-dwelling fish, would find suitable habitat throughout Lakes Erie and Ontario and along the shorelines of the other Lakes (Figure 10). In fact, this species became established in all five Great Lakes by 1998 (Rasmussen, 2002). Round gobies perch on rocks and other substrates in shallow areas, yet they have also been reported to flourish in a variety of habitat types (USGS, 2008a). Gobies also have a well developed sensory system that enhances their ability to detect water movement. This allows them to feed in complete darkness, giving them an advantage over other fish in the same habitat (Wisconsin Sea Grant, 2008). Zebra mussels may have facilitated the invasion of the round goby and other Eurasian species by providing an abundant food source (Ricciardi and MacIsaac, 2000). The distribution of the round goby around the inshore areas of the Black and Caspian seas indicates their potential for widespread occupation of inshore habitats with cover, especially plants, in the lower Great Lake, yet they can migrate to deeper water (50–60 m) in winter (Jude et al., 1992). The numbers of native fish species have declined in areas where the round goby has become abundant (Crossman et al., 1992). This species has been found to prey on darters, other small fish, and lake trout eggs and fry in laboratory experiments. They also may feed on eggs and fry of sculpins, darters, and logperch (Marsden and Jude, 1995) and have also been found to have a significant overlap in diet preference with many native fish species. They compete with rainbow darters (*Etheostoma caeruleum*), logperch (*Percina caprodes*), and northern madtoms (*Noturus stigmosus*) for small macroinvertebrates (French and Jude, 2001).

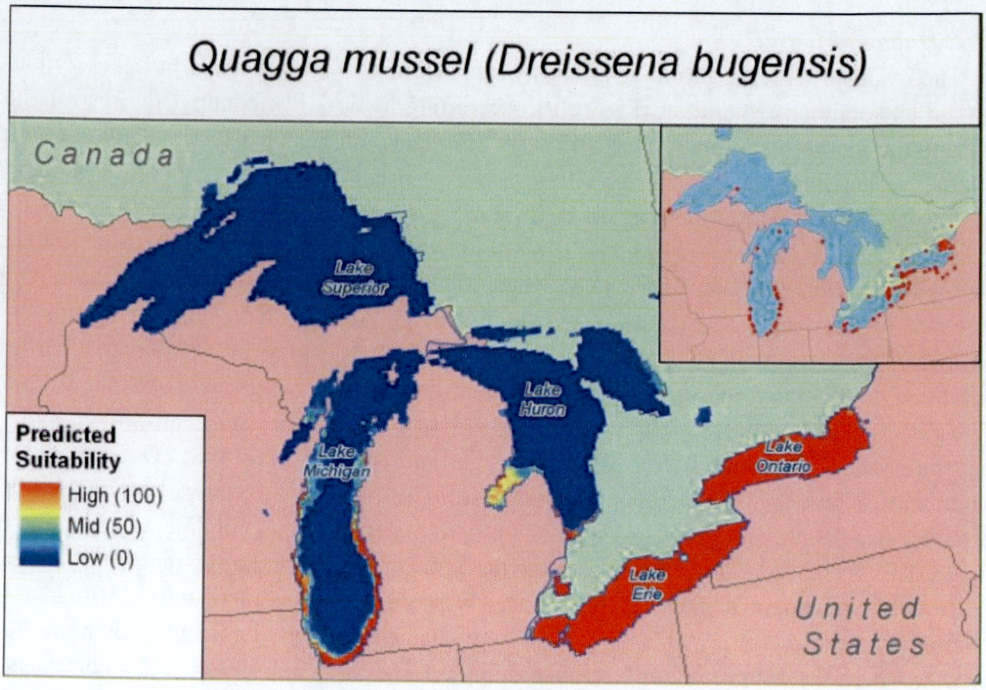

Figure 9. GARP-predicted habitat suitability of quagga mussel (*Dreissena bugensis*) in the Great Lakes. Inset map shows the locations where the species has been reported

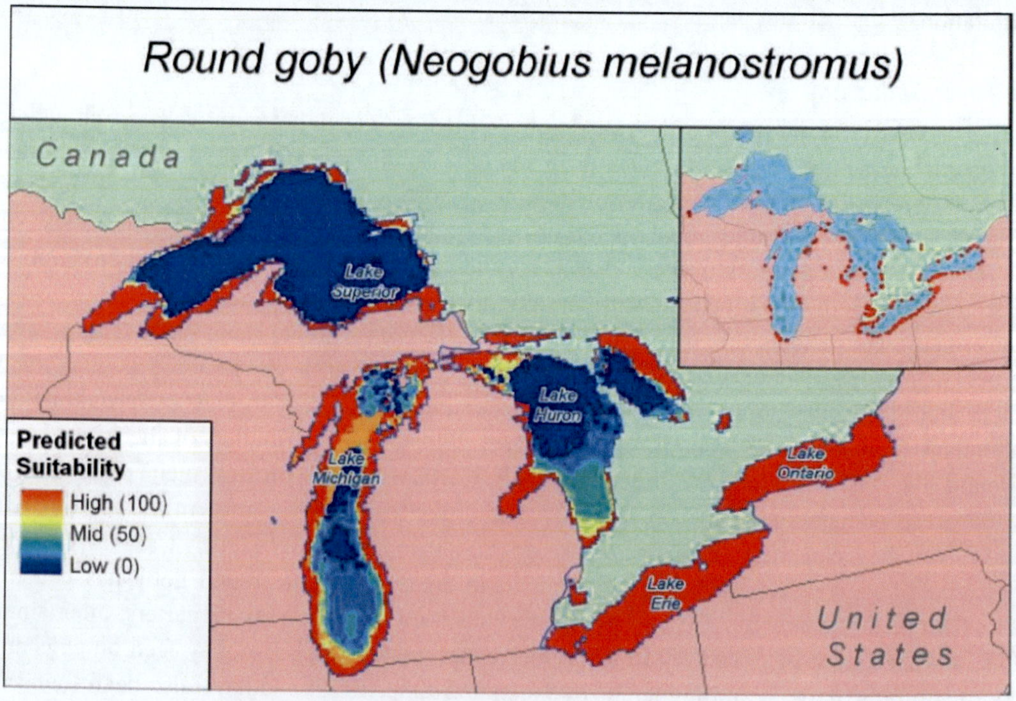

Figure 10. GARP-predicted habitat suitability of round goby (*Neogobius melanstromus*) in the Great Lakes. Inset map shows the locations where the species has been reported

4.2.4. Fishhook Waterflea

According to the GARP model, if transported to the Great Lakes, the fishhook waterflea, a free-swimming macroinvertebrate, would likely find suitable habitat throughout the region, except for the deeper waters of Lake Superior (Figure 11). The fishhook waterflea has been reported in Lakes Ontario, Michigan (USGS, 2008b), and Erie (University of Minnesota, 2006). The species is predicted to spread to the other Great Lakes, and, once established, it becomes difficult to eradicate (University of Minnesota, 2006). Unlike several of the other modeled species, population densities of the fishhook waterflea increase with distance from shore (IUCN, 2006), suggesting that this species may be able to occupy, given sufficient time, the entire region including the deeper waters of Lake Superior.

4.2.5. Zebra Mussel

The zebra mussel, a mollusk, has already invaded the shoreline areas of all five Great Lakes (Figure 12). The GARP model predicts the zebra model could potentially find suitable habitats throughout most of the Great Lakes region. Zebra mussels were first discovered in North America in 1988 in the Great Lakes. The first account of an established population came from Canadian waters of Lake St. Clair, a water body connecting Lake Huron and Lake Erie. By 1990, zebra mussels had been found in all the Great Lakes. The following year, zebra mussels escaped the Great Lakes basin and found their way into the Illinois and Hudson rivers. Zebra mussels are notorious for their biofouling capabilities by colonizing water supply pipes of hydroelectric and nuclear power plants, public water supply plants, and industrial facilities. Zebra mussels can have profound effects on the ecosystems they invade. They primarily consume phytoplankton, but other suspended material is filtered from the water column including bacteria, protozoans, zebra mussel veligers, other microzooplankton, and silt. Large populations of zebra mussels in the Great Lakes and Hudson River reduced the biomass of phytoplankton significantly following invasion. Diatom abundance declined over 80% and transparency, as measured by Secchi depth, increased by 100% during the first years of the invasion in Lake Erie (Holland, 1992). Zebra mussels represent one of the most important biological invasions into North America, having profoundly affected the science of Invasion Biology, public perception, and policy. Zebra mussels are described as poor O2 regulators, which may explain their low success rate in colonizing eutrophic lakes and the hypolimnion. Mellina and Rasmussen (1994) noted that calcium (Ca^{2+}) levels and water temperatures in the open waters of Lake Superior are too low for the zebra mussel. Zebra mussels require 10 mg/L of Ca^{2+} to initiate shell growth and 25 mg/L to maintain shell growth (USGS, 2008c). **Zebra mussels are generally within 2 to 7 m of the water surface (O'Neill,** 2004) but, on rare occasions, have been found at depths exceeding 90 m (Watkins et al., 2007). The depth limitation of the species should further restrict the maximum potential spread of the species to Lake Erie and to the shallower waters of the other four Great Lakes (NOAA, 2008). Competition with the quagga mussel also appears to limit zebra mussel spread. Zebra mussels were outcompeted and almost completely replaced by quagga mussels in Lake Ontario between 1995 and 2003 and this trend could occur in other Lakes (Watkins et al., 2007).

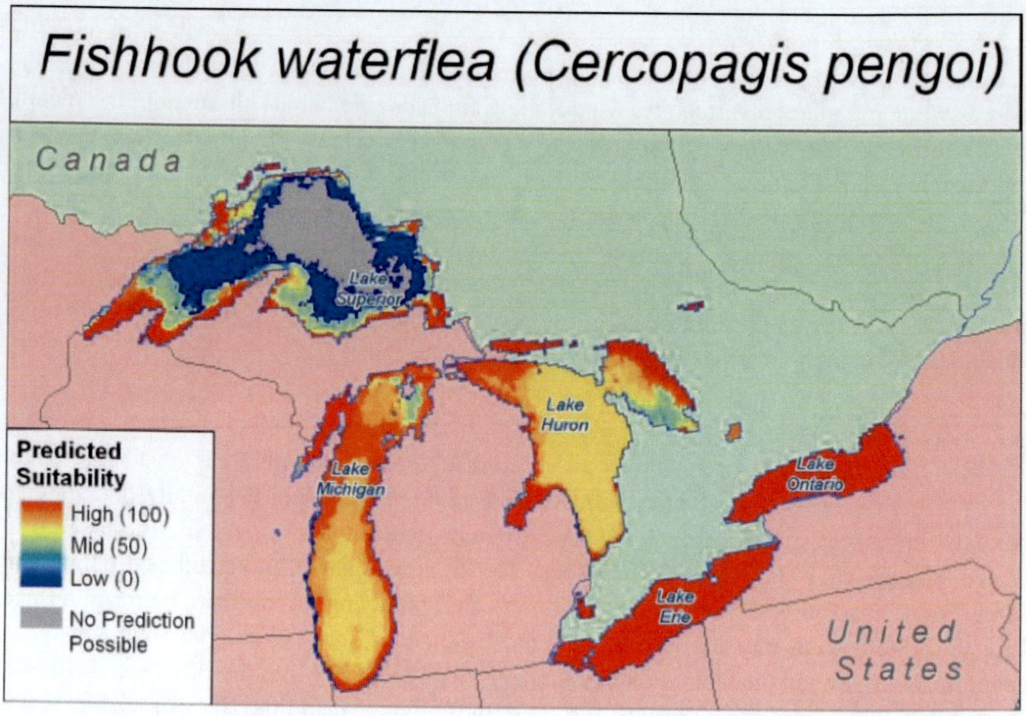

Figure 11. GARP-predicted habitat suitability of fishhook waterflea (*Cercopagis pengoi*) in the Great Lakes

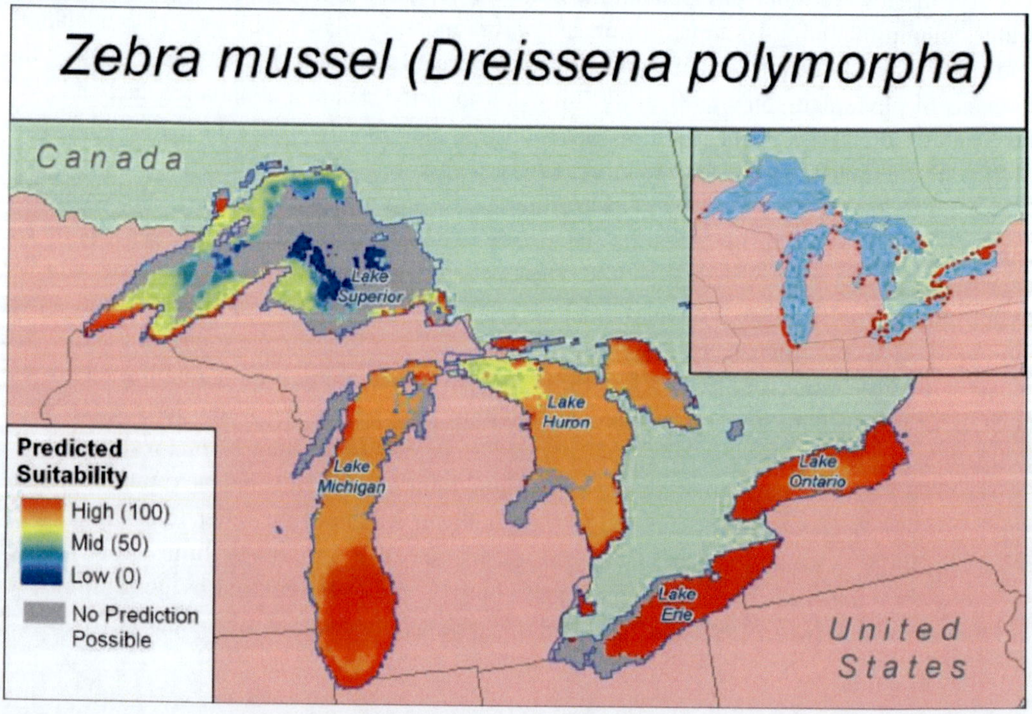

Figure 12. GARP-predicted habitat suitability of zebra mussel (*Dreissena polymorpha*) in the Great Lakes. Inset map shows the locations where the species has been reported

4.2.6. Ruffe

The ruffe, a small to medium-sized fish, has already invaded Lake Superior and GARP modeling predicts it will find suitable habitat almost everywhere in all five Great Lakes (Figure 13). GA RP models are not able to make a prediction about some of the deeper waters of Lake Superior. Established in the western portion of Lake Superior since about 1988 it has expanded in an easterly direction. It has now become the dominant species in the St. Louis River estuary (Leigh, 1998). Based on bottom trawl samples, ruffe make up an estimated 80% of fish abundances in the southwestern regions of Lake Superior (Leigh, 1998). Ruffe exhibit rapid growth and high reproductive output, and adapt to a wide range of habitat types; therefore, the species may pose a threat to native North American fish. There is much concern that ruffe may have a detrimental effect on the more desirable species in Lake Superior, such as yellow perch and walleye, by feeding on the young of these species or by competing for food (Fuller and Jacobs, 2008). Ruffe are often associated with bottom waters and can tolerate lacustrine and lotic systems and depths to 85 m (Sandlund et al., 1985). The species intolerance to deeper waters may limit its range of potential suitable habitat to Lake Erie, southern Lake Michigan, and the shallower waters of the other Great Lakes.

4.2.7. Monkey Goby

The GARP model predicts that the monkey goby, a member of the goby fish family, could find suitable habitat in most of Lake Erie and in some portions of Lake Ontario and Lake Huron (Figure 14). Predictions cannot be made for most of the Great Lakes because of data limitations. The monkey goby is closely related to the round goby. Currently, the monkey goby is confined to Eurasia but it has traveled up the Danube, Dnieper, and Volga Rivers from its native waterways and is becoming an invasive nuisance in these areas. Similar to other *Gobiidaes*, the monkey goby prefers shallow water and would likely not survive in deeper waters.

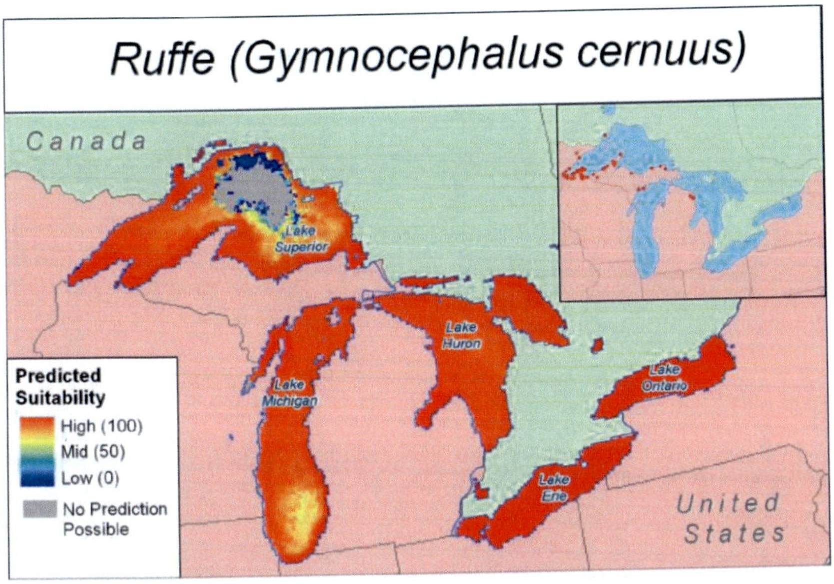

Figure 13. GARP-predicted habitat suitability of ruffe (*Gymnocephalus cernuus*) in the Great Lakes. Inset map shows the locations where the species has been reported

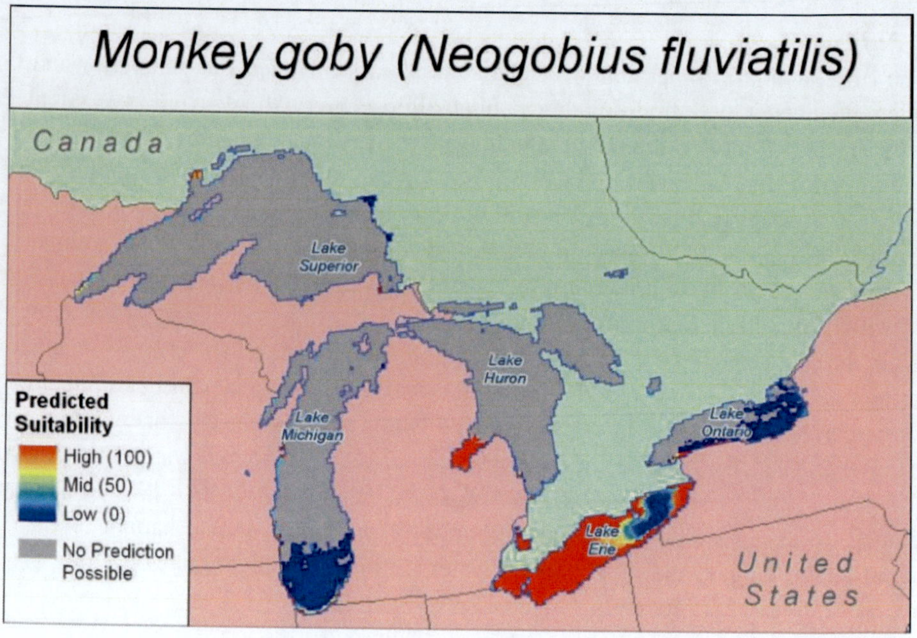

Figure 14. GARP-predicted habitat suitability of monkey goby (*Neogobius fluviatilis*) in the Great Lakes

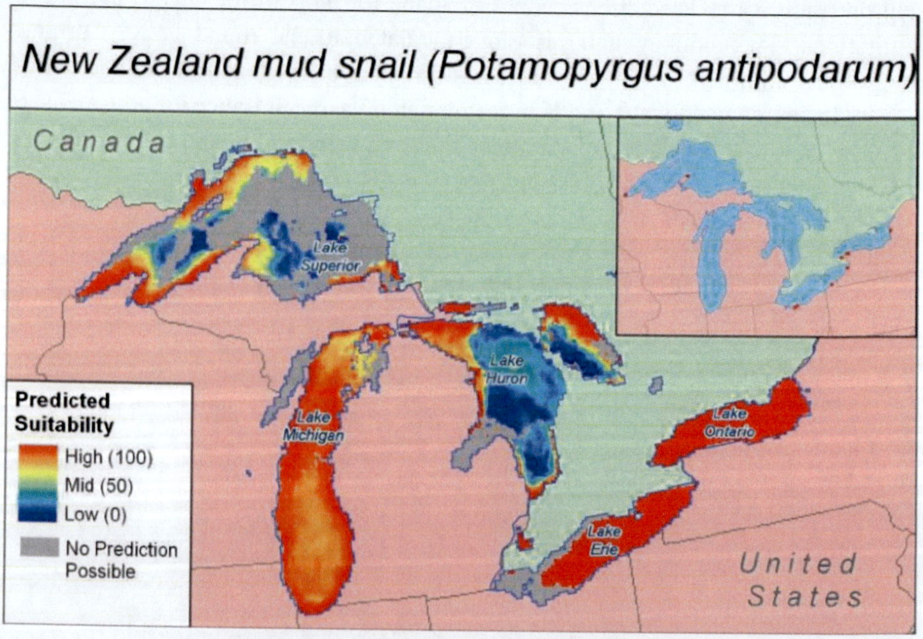

Figure 15. GARP predicted habitat suitability of New Zealand mud snail (*Potamopyrgus antipodarum*) in the Great Lakes. Inset map shows the locations where the species has been reported

4.2.8. New Zealand Mud Snail

The New Zealand mud snail, another mollusk, is predicted by GARP modeling to find suitable habitat in most if not all of Lakes, Erie, Ontario, and Michigan (Figure 15) and shorelines of Lakes Huron and Superior. It was first established in Lake Ontario in 1991 (Zaranko et al., 1997) and in Lake Erie in 2005 (Levri et al., 2007). It may also be established in Lake Superior, where some individuals were found in 2001 (Grigorovich et al., 2003b). Mud snail populations consist mostly of asexually reproducing females that are born with developing embryos in their reproductive system. This species can be found in all types of aquatic habitats from eutrophic mud bottom ponds to clear rocky streams. It can tolerate a wide range of water temperatures (except freezing), salinity, and turbidity in clean as well as degraded waters. They feed on dead and dying plant and animal material, algae, and bacteria. It can tolerate a broad range of ecological factors thus facilitating its further spread. In moist conditions, this snail can withstand short periods of desiccation. Since this snail is found at depths from 5 to 45 m (Levri et al., 2007) it is unlikely the species will survive in deeper waters.

4.2.9. Tubenose Goby

The tubenose goby, another member of the goby fish family, is predicted by the GARP model to become established in Lake Erie and the shoreline areas of the other Great Lakes (Figure 16). Predictions could not be made for most of the rest of the region. Their distribution around the inshore areas of the Black and Caspian Seas indicates their potential for widespread occupation of inshore habitats where cover, especially plants, occurs in the lower Great Lakes (Jude et al., 1992). Tubenose gobies have been shown to have a significant overlap in diet preference with rainbow darters, *Etheostoma caeruleum*, and may compete with these native fish for food (French and Jude, 2001). The usual habitat for this species is shallow bays, offshore banks, or flowing water of streams. However, the tubenose goby also can be found in ponds and canals overgrown with vegetation. Where current is strong, it hides under boulders. It is often found under stones or among weeds, to which it retreats rapidly if disturbed. Some individuals can be found at depths greater than 3 m in the sea. The preferred conditions probably restrict its probable range of suitable habitat to shallower waters.

4.2.10. Rudd

Already occurring in the Great Lakes with an unknown frequency at this time, significant portions of Lakes Erie and Ontario as well as portions of Superior and Michigan are prone to invasion by the rudd, a medium-sized, thick-bodied fish (Figure 17). The rudd's tolerance of a variety of habitats has likely contributed to its widespread distribution. In streams and rivers, this fish usually prefers long, slow pools and backwaters. The rudd can be expected to compete for invertebrate food sources with native fishes. In addition, being omnivorous, the rudd can shift its diet to plants, unlike most native fishes. Because rudd are fairly hardy, Nico et al. (2008) indicate that the fish will fare better than many native fishes in waters that are eutrophic or polluted. Predictions cannot be made about the habitat suitability for rudd in parts of Lake Superior, but, given the species preference for littoral waters, it is unlikely the rudd would find suitable habitat in the deeper regions of all the lakes.

4.2.11. Corophium Curvispinum (an Amphipod)

According to the GARP model, almost all of Lake Erie and the southern shores of Lakes Ontario, Huron, and Michigan are prone to invasion by the amphipod *Corophium curvispinum*. Predictions for the other locations in the Great Lakes were not possible due to limited data (Figure 18). This amphipod builds tubes on firm surfaces such as rocks, wood, submerged vegetation, or bivalve shells on otherwise sandy or muddy substrata in shallow waters (Frammandearter, 2008). *C. curvispinum* prefers rivers, estuaries, and other areas with brackish water, but it can also tolerate freshwater environments.

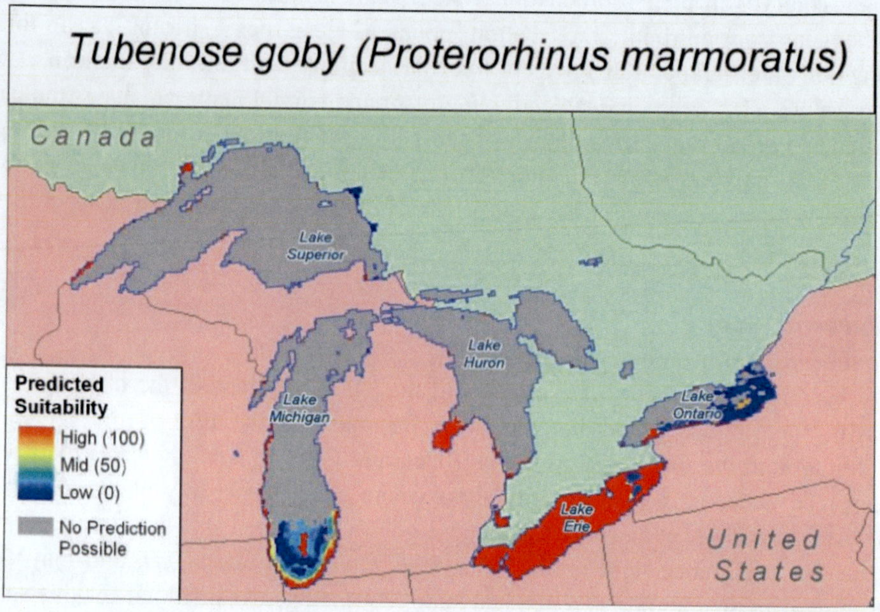

Figure 16. GARP-predicted habitat suitability of tubenose goby (*Proterorhinus marmoratus*) in the Great Lakes

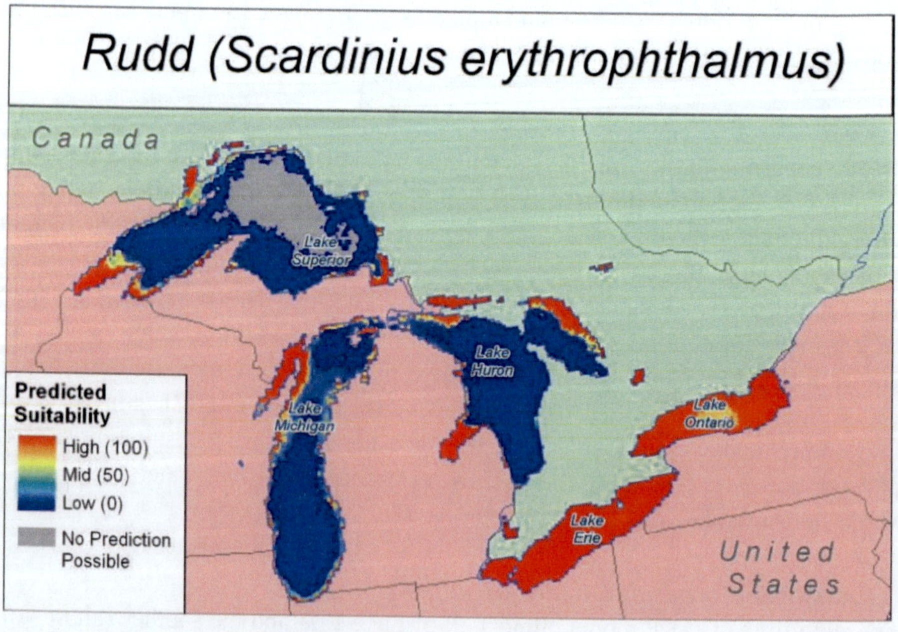

Figure 17. GARP-predicted habitat suitability of rudd (*Scardinius erythrophthalmus*) in the Great Lakes

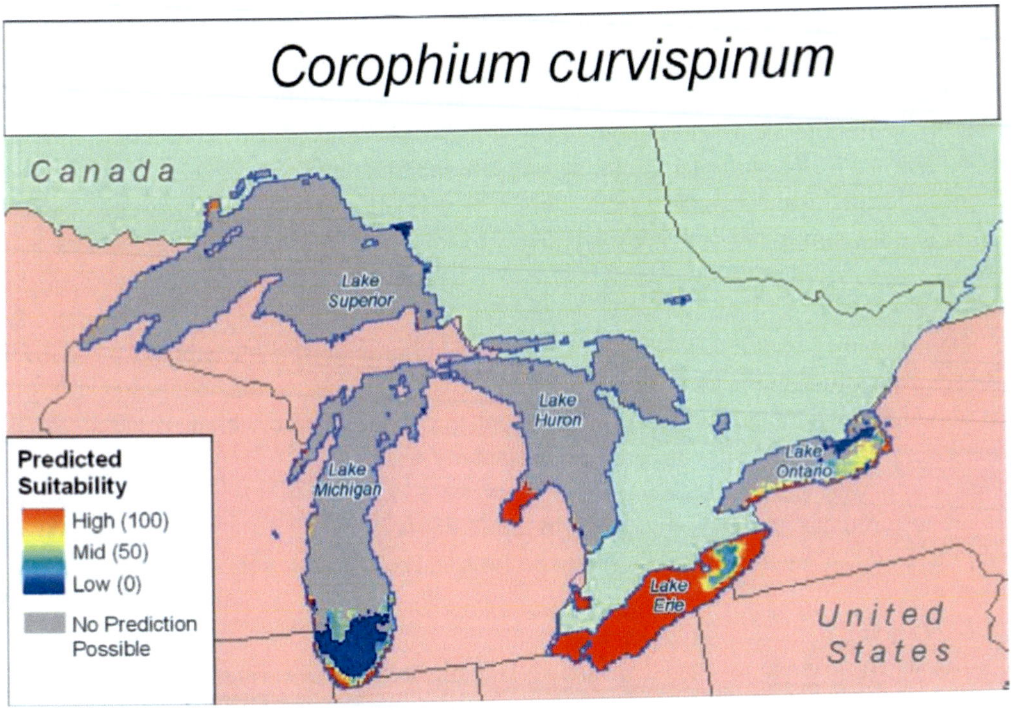

Figure 18. GARP-predicted habitat suitability of *Corophium curvispinum* (no common name reported) in the Great Lakes

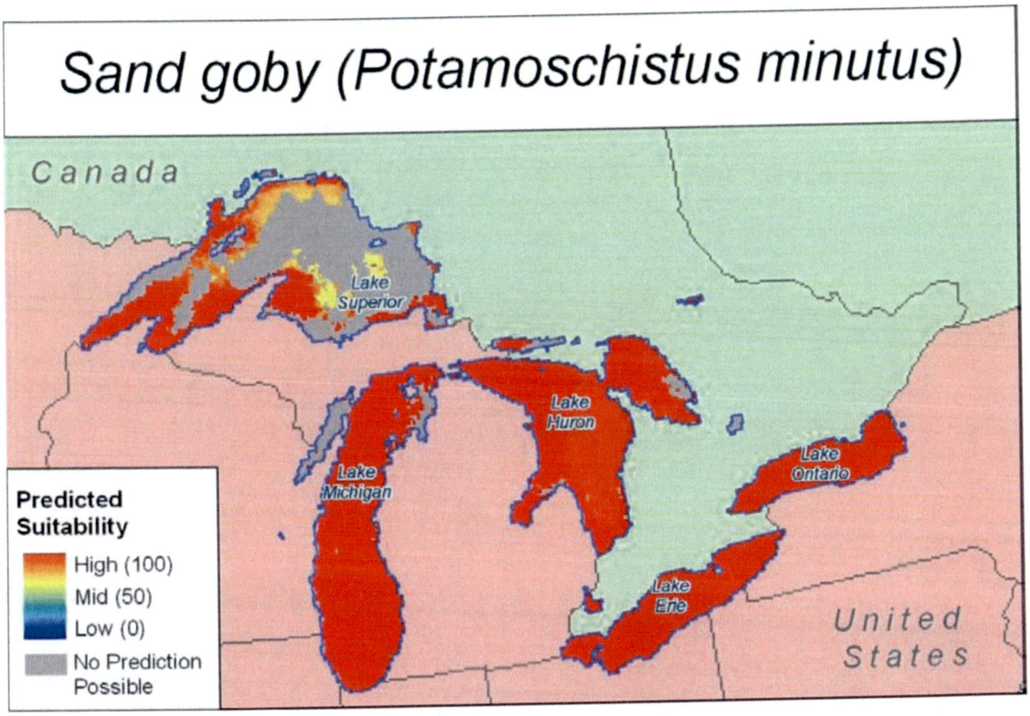

Figure 19. GARP-predicted habitat suitability of sand goby (*Potamoschistus minutus*) in the Great Lakes

4.2.12. Sand Goby

The sand goby is predicted by GARP modeling to find suitable habitat almost everywhere in all five Great Lakes (Figure 19). This occasionally schooling species occurs primarily in inshore sandy and muddy areas (Froese and Pauly, 2008). The sand goby is a coastal goby of European waters from the Baltic to the Mediterranean Sea and can grow up to 10 cm in length. Some variation from the GARP modeling prediction is expected because similar to other gobys, the sand goby is unlikely to find suitable habitat in deeper waters.

4.2.13. Roach

GARP predicts that Lakes Erie and Ontario would provide suitable habitat for the roach, a medium-sized fish in the carp family. Most of the other regions would be unsuitable (Figure 20) although predictions cannot be made about the suitability of habitat of parts of Lake Superior. Adults inhabit slow-flowing or still muddy waters and are abundant in their native rivers, lakes, canals, and reservoirs. Brackish water populations in the Baltic and the Black Sea are anadromous and they are known to thrive in poor quality, even polluted water (Nico and Fuller, 2008). As omnivores, they feed on insects, crustaceans, mollusks, and plants.

4.2.14. Tench

The tench, a medium-sized fish already established in many rivers within the United States, is likely to find suitable habitat in most of Lake Erie and small portions of Lake Ontario and Lake Huron (Figure 21) according to GARP models. The diet consists mainly of aquatic insect larvae and mollusks. Nico and Fuller (2008) considered it a potential competitor for food with sport fishes and native cyprinids and noted that the species is known to stir up bottom sediments, possibly affecting water quality, similar to the common carp.

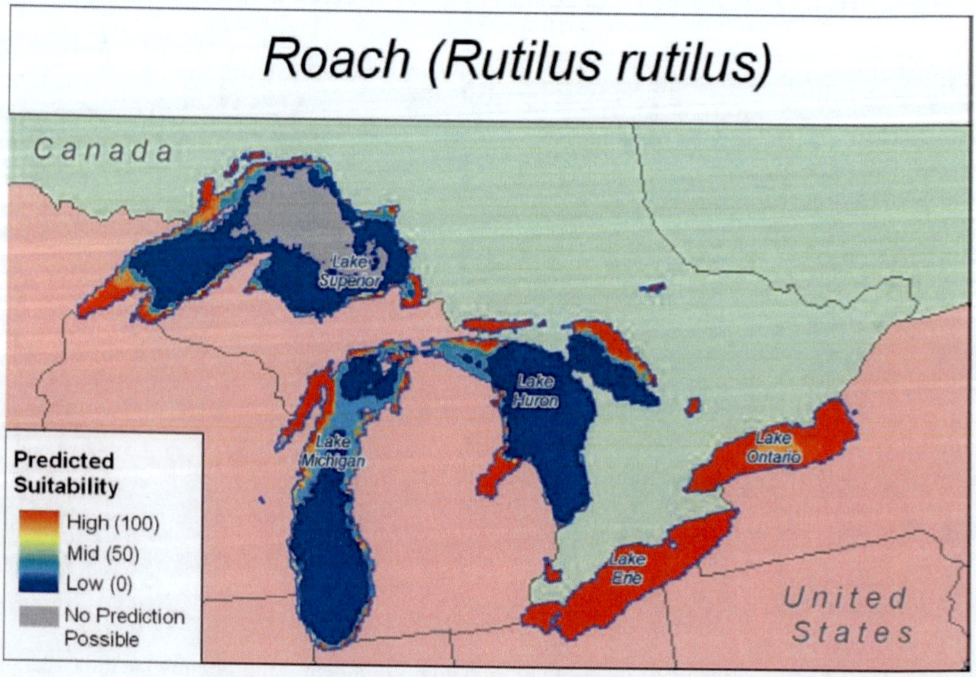

Figure 20. GARP-predicted habitat suitability of roach (*Rutilus rutilus*) in the Great Lakes

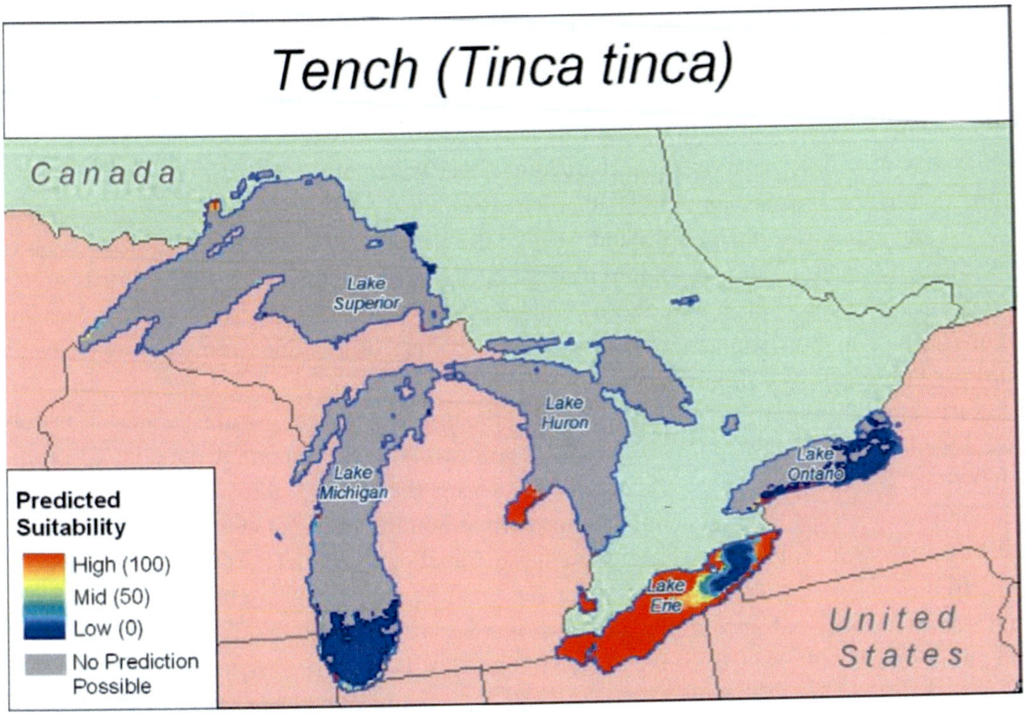

Figure 21. GARP-predicted habitat suitability of tench (*Tinca tinca*) in the Great Lakes

4.3. Vessel Traffic and Great Lakes Ports

The U.S. Great Lakes receive substantial vessel traffic from around the world because of the commodities shipped in and out of the area. The second phase of this study was to better understand whether there is sufficient numbers of propagules (e.g., larvae, seeds, spores, adults) entering the Great Lakes for species to become established. As described in the methods section, we used vessel traffic and ballast water discharges as a surrogate for propagule pressure since no data exists that actually measures the number of propagules released from discharges. By analyzing ballast water discharges, we were able to identify those ports that are at greatest risk.

4.3.1. Analysis of Vessels with Ballast on Board (BOB)

In accordance with U.S. Coast Guard (USCG) regulations, ballast water from transoceanic vessels with ballast on board is exchanged at sea prior to entering the St. Lawrence Seaway. Despite the ballast water exchange (BWE), some ballast water and residue may remain and NIS may survive in the ballast tank and then potentially be released when the ballast water is discharged. We evaluated ballast water discharges (BWD) into U.S. Great Lakes ports from vessels entering the Seaway whose original source of ballast water (i.e., the ballast in the tank prior to open ocean exchange) was taken-on from areas outside of the Great Lakes. To interpret these results one must consider that transoceanic vessels carry multiple ballast tanks and each tank may have a different history of ballast water source, exchange, and discharge. Therefore, each ballast tank discharge was counted as a BWD. A transoceanic vessel may carry over 20,000 metric tons of ballast water and have as many as 20 ballast tanks, implying one vessel-trip could have up to 20 ballast discharges at any one Great Lakes port.

Our study found that in the period of 2006–2007, 618 ballast tanks and 382 thousand metric tons of ballast water were discharged at Great Lakes ports from 107 different vessels. From a global perspective, the BWDs that were evaluated and could be linked to a particular port usually came from the eastern and western areas of the northern Atlantic Ocean (Figure 22). The Gulf of St. Lawrence region, near the St. Lawrence Seaway was the original source of ballast water for over 1/3 of the 618 discharges (Figure 23). Western Europe was the second-most common source, with most of the ballast originating from the southeastern portion of the North Sea (Figure 24).

Fifty-eight different foreign ports provided the original source of ballast water ultimately discharged at Great Lakes ports. Figure 25 identifies the most important ports based on the number of vessels leaving these ports and entering the Great Lakes. However, the most common source of ballast water was obtained while in transit and not at any particular port of call. The ports of Antwerpen, Belgium; Puerto Cabello, Venezuela; Haraholmen, Sweden; and Bremen, Germany are responsible for the greatest number of discharges, not including those ports near the entrance to the St. Lawrence Seaway (Figure 25). These ports, however, are not necessarily the source of the greatest volume of BWD. For instance, Haraholmen was ranked fourth among nonNorth American ports in terms of number of tanks discharged, but was ranked eighth in terms of metric tons of BWD (Figure 25 and Appendix E, Table E-1). In this study it was rare to find more than one vessel originating from the 58 foreign ports (Appendix E, Table E-1). Only six ports from outside North America were the source of two or more vessels included in this analysis.

Duluth received more than twice the BWDs and twice the volume of ballast water as any other Great Lake port in 2006–2007 (Figure 26). The ports of Toledo, Superior, Green Bay, Gary, and Milwaukee, also received over 10,000 metric tons of ballast water (Figure 26 and Appendix E, Table E-2). Appendix E, Table E-3 provides detailed information for those vessels with ballast on board discharging to Great Lakes ports in 2006–2007.

Figure 22. Location of the original source of ballast water taken-on prior to ballast water exchange and discharge in the U.S. Great Lakes. The area of each green circle is proportional to the number of ballast tank discharges

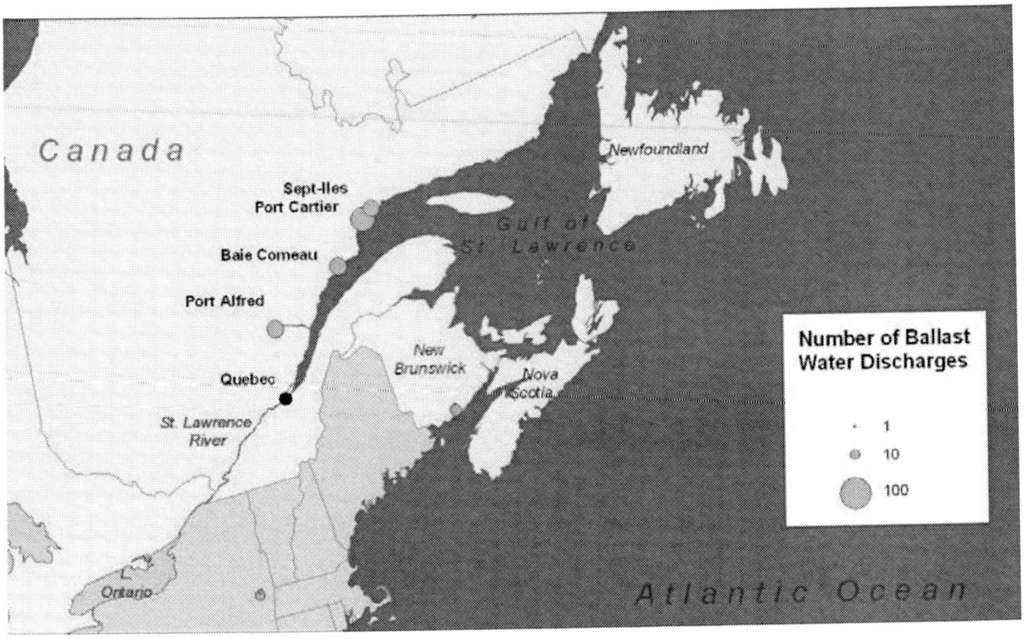

Figure 23. Location of the source of ballast water taken-on from Canadian ports in or near the Gulf of St. Lawrence prior to ballast water exchange and discharges in the U.S. Great Lakes. The area of each green circle is proportional to the number of ballast tank discharges

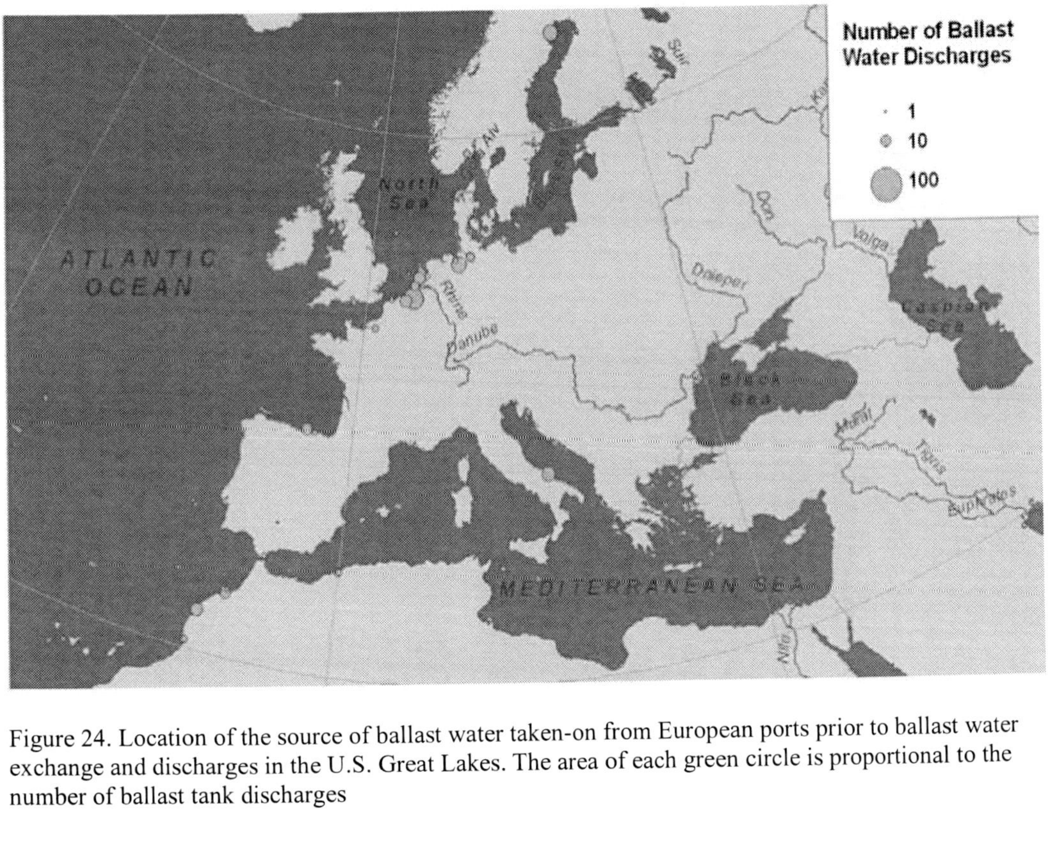

Figure 24. Location of the source of ballast water taken-on from European ports prior to ballast water exchange and discharges in the U.S. Great Lakes. The area of each green circle is proportional to the number of ballast tank discharges

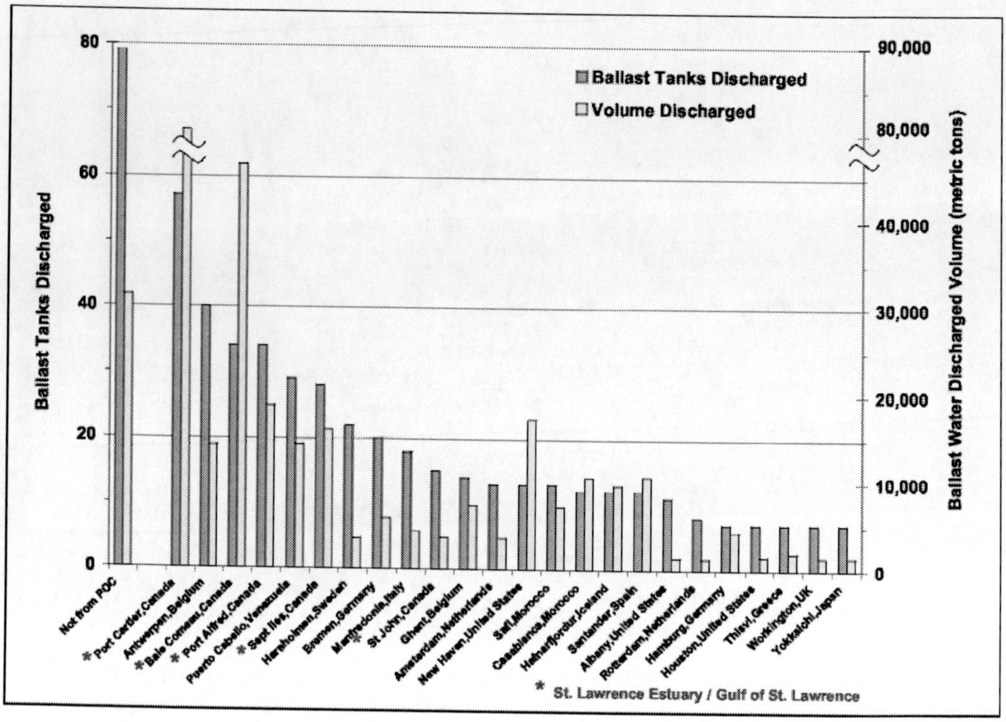

Figure 25. Frequency, volume, and original source of ballast water (prior to ballast water exchange) discharged into U.S. Great Lakes ports, from sources outside the Great Lakes

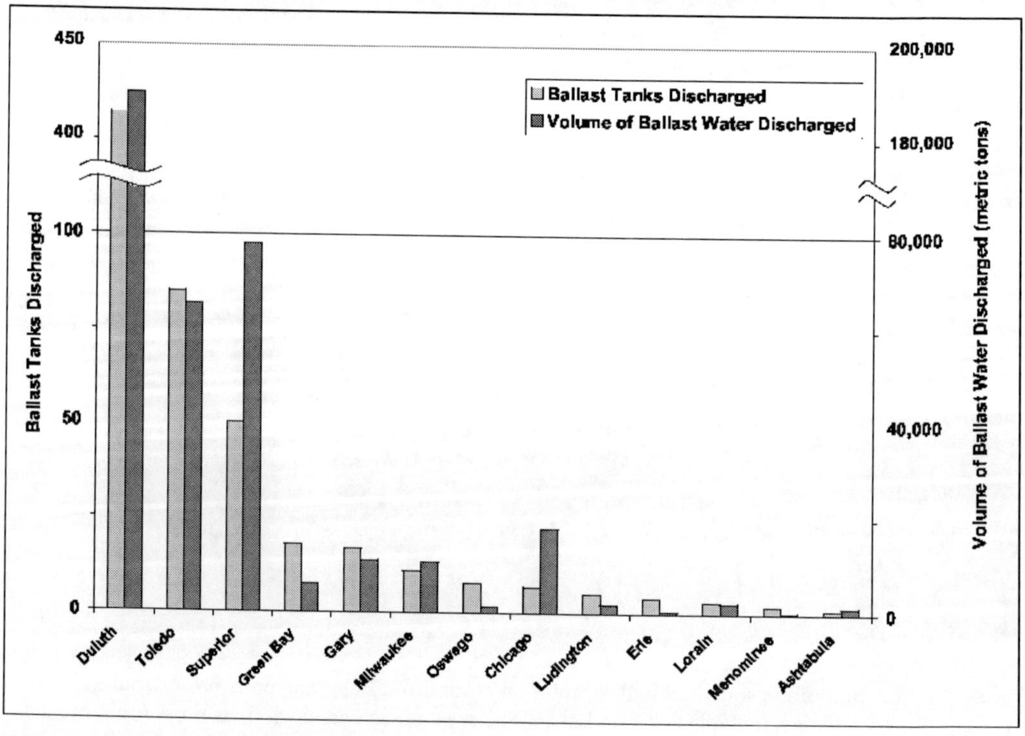

Figure 26. Frequency and volume of ballast water discharges (after ballast water exchange at sea) from ballast on board vessels, when the original source of ballast water came from outside the Great Lakes

4.3.2. Analysis of Vessels with No Ballast on Board (NOBOB)

Some vessels enter the St. Lawrence Seaway with no ballast on board but may have organisms that remain and survive in the residual material left in the ballast tanks, and are referred to as NOBOB-RM vessels. These NOBOB-RM vessels can then take-on ballast water in the Seaway (most likely when cargo is off-loaded) and later discharge the ballast water along with residual materials at a Great Lakes port. We combined the 2006 NVMC data with the 2006 NBIC data to identify these types of vessels. There were 1,730 discharges of ballast water at Great Lakes ports from NOBOB-RM vessels in 2006. This is substantially more than the discharges from vessels with ballast on board, supporting the notion that NOBOB vessels may pose a much greater risk. The distribution of the potential sources of ballast water from NOBOB-RM vessels is somewhat similar to the vessels with ballast on board. Over half of the last five ports of call by these vessels were in southeastern Canada with Western Europe the second most common source of ballast water (Figure 27 and Appendix E, Table E-4).

Some of the foreign ports of origin are different between BOB and NOBOB vessels. For example, ten NOBOB-RM vessels included a stop at Europa Point, Gibraltar as one of the last five ports of call before discharging ballast water into the Great Lakes. Other vessels stopped at Riga, Latvia; and Santos, Brazil (Figure 27 and Appendix E, Table E-4). Yet, we did not find any vessels with ballast on board stopping at these ports. Some ports are visited by both types of vessels, especially Sept Iles, Canada; and Ijmuiden, Netherlands (Figure 27 and Appendix E, Table E-4).

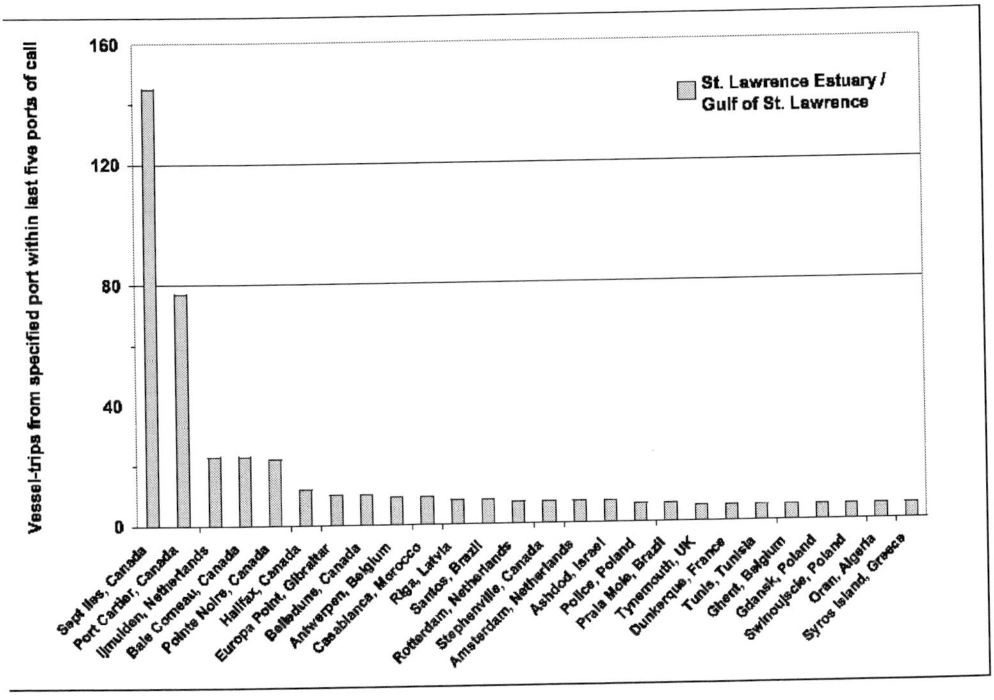

Figure 27. Possible source locations of residual materials discharged from vessels that entered the St. Lawrence Seaway with no ballast on board, based on last five ports of call in 2006. After entering the Seaway these vessels picked up ballast water and discharged the ballast water (along with residual materials) at a U.S. Great Lakes port.

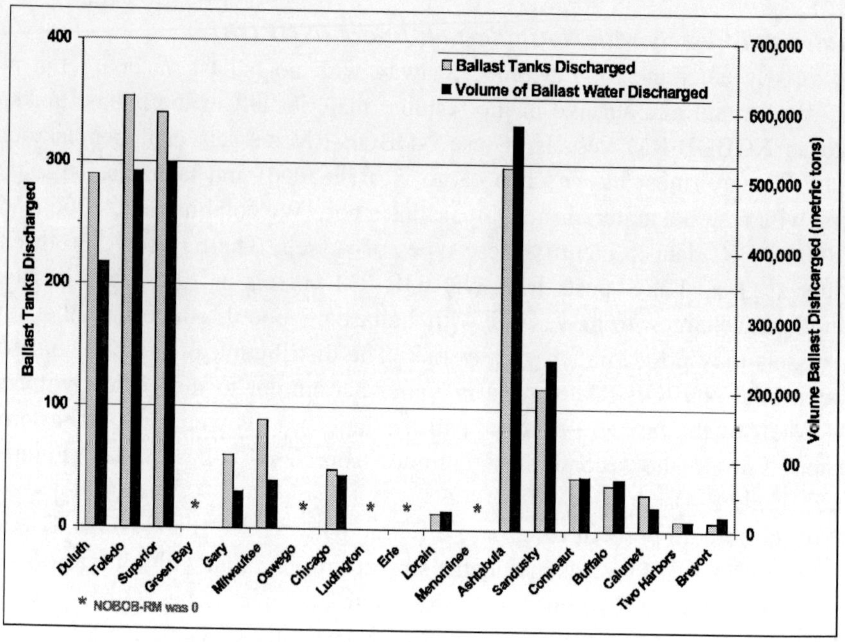

Figure 28. Frequency and volume of discharges in 2006 from NOBOB-RM vessels. NOBOB-RM vessels are vessels that entered the Seaway without ballast on board, picked up ballast water in the Great Lakes, and then discharged the ballast water along with residual material at a Great Lake port. NOBOB-RM vessels must also have visited a port outside the Great Lakes during one of the last five ports of call. Ports listed from left to right according to the results shown in 26 to facilitate comparison.

Several of the Great Lakes ports, including Duluth, Toledo, and Superior, receive ballast discharges from both NOBOB-RM vessels and vessels with ballast on board. Most Great Lakes ports received far more ballast discharges from NOBOB-RM vessels than BOB vessels in 2006. Several ports receive most of their ballast water from NOBOB-RM vessels, including Sandusky, Conneaut, Buffalo, and Calumet as shown in Figure 28 and Appendix E, Table E-5. Ashtabula was the extreme case, receiving 297 discharges from NOBOB-RM vessels (Figure 28) and only one discharge from a BOB vessel (Figure 26). When both vessel types are considered, the Great Lakes ports at greatest risk of receiving sufficient propagule pressure to facilitate invasion are Duluth, MN; Superior and Milwaukee, WI; Toledo, Ashtabula, and Sandusky, OH; Gary, IN; and Chicago, IL.

5. Discussion

The Great Lakes system has been adversely affected by invasive species. Preventing the transport of these species to the Great Lakes from outside the system is the best way to avert potential ecological and economic impacts. Our analysis of ballast water discharges using vessel traffic data, evaluating similar habitats using the Genetic Algorithm for Rule-Set Production (GARP) niche model, and a literature review indicate that invasions are likely to occur over the next decade or so. If it is not possible to eliminate the transport of nonindigenous species (NIS) to the Great Lakes, the next best alternative is to monitor for the arrival of potentially invasive species and to control their spread as soon as they arrive.

Table 3. Composite Results for 14 Species Modeled Using GARP

Species/common name	Summary of invasion potential
Species already established in the Great Lakes and potential for spread to all five Great Lakes	
Neogobius melanostromus—round goby	Already spread to all five Great Lakes, with large populations in Lakes Erie and Ontario. Likely to find suitable habitat throughout Lake Erie and in all Great Lakes waters at depths less than 60 m.
Potamopyrgus antipodearum—New Zealand mud snail	Already occurs in isolated areas of Lakes Erie, Ontario, and Superior. Likely to find all shallower waters (<50 m depth) as suitable habitat. High spread potential.
Dreissena bugensis—quagga mussel	Already found in all five Great Lakes, with large populations established in Lakes Erie and Ontario. The only possible identified limitation for spread is a species depth limitation which is questionable and currently appears to be as deep as 200 m.
Dreissena polymerpha— zebra mussel	Already occurs in all five Great Lakes. Likely to find suitable habitat in most of Lake Erie and portions of other lakes where water depth is less than 60 m. May be outcompeted by the quagga mussel.
Gymnocephalus cernuus— ruffe	Already found in Lakes Superior, Michigan, and Huron. The species is probably capable of colonizing most areas within the Great Lakes where water depth is less than 85 m.
Species that may invade at least parts of all five Great Lakes	
Alosa aestivalis—blueback herring	Models predict it could find the entire region as suitable habitat, except possibly the deeper waters of Lake Superior.
Pomatoschistus minutus— sand goby	It is likely this species would find all shallower waters as suitable habitat.
Rutilus rutilus—roach	Already reported in Lakes Erie and Ontario. Predicted to find suitable habitat throughout these lakes, and probably into other shoreline areas.
Scardinius erythrophthalmus—rudd	Predicted to find suitable habitat throughout Lake Erie and into the shallower waters of the other four Great Lakes.
Cercopagis pengoi—Established fishhook waterflea	Established in Lake Ontario and reported in Lakes Erie and Michigan. Predicted to find suitable habitat throughout the region. Densities increase in deeper waters.
Tinca tinca—tench	Found currently in St. Lawrence River. Potential to spread to shallower waters of most of Lake Erie, and to isolated portions of the other Lakes. Tench can spread rapidly once established.
Proterorhinus marmoratus—tubenose goby	Already reported as present in Lake St. Clair and western Lake Erie. May be able to occupy all shallow waters of all five Great Lakes.
Corophium curvispi num— (an amphipod)	Capable of invading Lake Erie and shallower waters. Not enough data is available to predict if it can find suitable habitat elsewhere.
Neogobius fluviatilis— monkey goby	May be capable of inhabiting shallower waters of all five Great Lakes.

Since we began our investigation, additional ballast-water control measures have been implemented. Beginning with the 2008 navigation season, all vessels must either undergo ballast water exchange (BWE) or flushing before entering the St. Lawrence Seaway (73 FR 37, p. 9,950), even those vessels that heretofore were declared as having no ballast on board. However, even with the more extensive requirements, additional NIS may still reach the Great Lakes. Some saltwater tolerant species may survive the BWE or flushing, and other vectors (e.g., hull fouling organisms) continue to pose a threat. This chapter provides information that may help resource managers prioritize monitoring efforts by identifying potential invaders and ports at risk.

The National Research Council recommends that a binational (United States and Canada) science-based surveillance program be established to monitor for aquatic NIS and that the program involve dedicated lake teams, as well as academic researchers, resource managers, and local citizens groups (NRC, 2008). Since early detection and rapid response is a priority of the National Invasive Species Council (NISC, 2007), the ports and species we identified could be used to structure an early warning and detection system to help evaluate the effectiveness of ballast exchange regulations and practices.

5.1. Predicting the Spread of Species

This study identified 30 potentially invasive species with medium or high risk for spreading to the Great Lakes and causing ecological impacts and another 28 potentially invasive species that have already become established in one or more of the Great Lakes (see Appendix B). Habitat suitability maps are provided along with a summary of invasion potential for 14 modeled species. All of the modeled species are predicted to have the capability to colonize Lake Erie and the shallower waters of the other lakes. Several species may be able to colonize the entire Great Lakes region. Literature regarding the species environmental tolerances reveals that the predominant limitation to the spread of several modeled species is their tolerance to water depth. However, managers need to recognize that when NIS are transported to a new environment, species-tolerance "surprises" can occur. For instance, the quagga mussel has been found at deeper depths in the Great Lakes than in its native range (Watkins et al., 2007).

Table 3 summarizes the habitat suitability and current status for the 14 modeled species. The modeled species are categorized into two groups: (1) N IS already established in the Great Lakes and having the potential to spread to at least parts of all five lakes; and (2) NIS, not yet established but with the potential to invade the Great Lakes.

5.2. Potential Monitoring Sites Based on Vessel Traffic

The source of most of the ballast water discharged into the Great Lakes came from 58 different ports located predominantly in Canada and Western Europe, thereby complicating surveillance programs. If just a few foreign ports were the original source of ballast water (prior to exchange) then programs could focus on species found at those foreign ports. The six ports which received the most ballast water from vessels with ballast on board in 2006–2007, in rank order, were Duluth, MN; Toledo, OH; Superior, WI; Green Bay, WI; Gary, IN; and Milwaukee, WI (see Figure 26 and Appendix E, Table E-2). The first three ports, Duluth, Toledo, and Superior, account for 86% of the total volume of ballast water discharged into the Great Lakes. There was no evidence of a frequent, repeated connection from any specific foreign port to a specific port within the Great Lakes. For instance, 11 different vessels discharged ballast water in Toledo in 2006–2007 (see Appendix E, Table E-3). If all 11 vessels obtained ballast water from a single foreign port than monitoring could be targeted for those species occurring at that particular port. Unfortunately, the 11 vessels discharging ballast water in Toledo took-on ballast from 10 different foreign ports.

Invasive species can also be transported to the Great Lakes via vessels with no ballast on board but with residue left in the tanks (NOBOB-RM vessels). A different set of foreign ports were found to be the source of ballast water from these vessels (see Figure 27), although ports in Canada and Western Europe predominated. Consistent with the vessels with ballast on board, it was rare to find a frequent connection between particular Great Lakes ports and foreign ports. The ports receiving the greatest volume of ballast water from NOBOB-RM vessels are Duluth, MN, Toledo, OH, and Superior, WI accounting for 54% of the total volume discharged from these vessels in 2006–2007. Ashtabula and Sandusky, OH receive 32% of the ballast water released from NOBOB-RM vessels.

Since it is unknown which type of vessel (ballast on board [BOB] or NOBOB-RM) is more likely to transport NIS, ports receiving ballast water from either type of vessel are presumed to be at risk. In order to recommend ports for monitoring we cannot just consider the transport potential, we also need to consider the potential to find suitable habitat. The results of GARP modeling and the literature review reveal that Lake Erie and shallower portions of the other Lakes provide the most favorable habitat for the modeled species, and that the deeper portions of Lake Superior are less hospitable to species invasions (Grigorovich et al., 2003b). However, the shallower portions of Lake Superior, especially the Duluth-Superior harbor, are at greater risk for invasion.

Assuming the observed vessel traffic and ballast-water discharge information for 2006 and 2007 is representative, the port of greatest concern for receiving sufficient propagules and providing the most suitable habitat is Toledo, OH. Toledo is located on Lake Erie, a region that the GARP model predicted would have a high chance of providing suitable habitat for the modeled species. Other ports of elevated concern for receiving sufficient propagules and offering suitable habitat are Gary, IN; Milwaukee, WI; Chicago, IL; and Ashtabula and Sandusky, OH. Ports with high transport potential but generally low habitat suitability are Duluth, MN and Superior, WI. The spread of invasive species from beyond the Duluth-Superior harbor may be limited by the colder and deeper waters in the main portion of Lake Superior. Yet, since inter-lake transport can occur to other ports, Duluth and Superior also warrant a monitoring program. Managers may wish to emphasize detection programs at the ports of concern that were identified and may wish to focus on the list of 58 potentially invasive species with a moderate or strong chance to invade and cause ecological or economic impacts. For the 14 modeled species, the focus can be narrowed based on the summary shown in Table 3.

Given the new regulations which require all vessels entering the Seaway to undergo either ballast water exchange or flushing at sea, additional research on the tolerance of invasive species to saltwater would enable managers and scientists to better focus monitoring activities on those species that are likely to survive salt water flushing. Subsequent analyses of the NBIC database is recommended to determine if the 2006–2007 data are indeed representative.

In summary, we have provided a list of NIS of concern, predicted locations that would provide suitable habitat for 14 modeled species, identified those U.S. Great Lakes ports receiving the most ballast water from sources originating from outside the Great Lakes, and predicted the ports most at risk of invasion. Our findings support the need for detection and monitoring efforts at those ports believed to be at greatest risk. This study also demonstrates the importance of understanding invasion biology by evaluating the two most important predictors of invasion, as suggested by Williamson (1996): propagule pressure and suitable habitat. Further, this may be the first time that remote sensing data were used in conjunction with GARP to predict the spread of aquatic invasive species.

REFERENCES

Anderson, R. P., Laverde, M. & Peterson, A. T. (2002). Geographical distributions of spiny pocket mice in South America: Insights from predictive models. *Global Ecology and Biogeography, 11,*131-141.

Anderson, R.P., Lew, D. & Peterson, A. T. (2003). Evaluating predictive models of species distributions: criteria for selecting optimal models. *Ecol Mod, 162(3),*:211–232.

Bailey, S. A., Reid, D. F. & Colautti, R.I. et al. (2005). Management options for control of nonindigenous species in the Great Lakes. *J Great Lakes L Sci and Policy, 5*,101–112.

Buchan, L. J. & Padilla, D. K. (2000). Predicting the likelihood of Eurasian watermilfoil presence in lakes, a macrophyte monitoring tool. *Ecol Appl, 10,*1442–1445.

Carlton, J. T., Reid, D. M. & van Leeuwen, H. (1995). Shipping study- the role of shipping in the introduction of non-indigenous aquatic organisms to the coastal waters of the United States (other than the Great Lakes) and an analysis of control options. In: *The National Biological Invasions Shipping Study*. United States Coast Guard and the National Sea Grant Program, Washington, D.C., USA.

Claxton, W. T., Wilson, A. B. & Mackie, G. L., et al. (1998). A genetic and morphological comparison of shallow- and deep-water populations of the introduced dreissenid bivalve *Dreissena bugensis*. *Can J Zool, 76(7),*1269–1276.

Cohen, A. N. (2007). Potential distribution of zebra mussels (*Dreissena polymorpha*) and quagga mussels (*Dreissena bugensis)* in California phase 1 report. California Department of Fish and Game. Available on-line at http://www.dfg.ca.gov/invasives/quaggamussel/docs/Potential DistributionZebraQuaggaInCA.pdf (accessed March 2008).

Colautti, R.I. Niimi, A. J. & Van Overdijk, C. D. A. et al. (2003). Spatial and temporal analysis of transoceanic shipping vectors to the Great Lakes. In: Ruiz, GM; Carlton, JT; eds. *Invasive species: vectors and management strategies*. Washington, DC: Island Press; pp. 227–246.

Costello, C., Drake, J. & Lodge, D. (2007a). Evaluating an invasive species policy: ballast water exchange in the Great Lakes. *Ecol Appl, 17(3)*, 655–662.

Cox, G. W. (2004). *Alien species and evolution: the evolutionary ecology of exotic plants, animals, microbes, and interacting native species*. Washington, DC: Island Press.

Crossman, E. J., Holm, E. & Cholmondeley, R. et al. (1992). First record for Canada of the rudd, Scardinius erythrophthalmus, and notes on the introduced round goby, *Neogobius melanostomus. Canadian Field-Naturalist, 106(2),* 206–209.

Drake, J. M. & Bossenbroek, J. M. (2004). The potential distribution of zebra mussels in the United States. *Biol Sci, 54,* 931–941.

Duggan, I. C., Bailey, S. A. & Colautti, R. I. et al. (2003). Biological invasions in Lake Ontario: past, present and future. In: Munawar, M.., ed. State of Lake Ontario (SOLO) – past, present and future. Ecovision World Monogr. Aquat Ecosyst Health Manage, pp. 1–17.

Duggan, I. C., Van Overdijk, D. A. & Bailey, S. A. et al. (2005). Invertebrates associated with residual ballast water and sediments of cargo-carrying ships entering the Great Lakes. *Can J Fish Aquat Sci, 62(11),* 2463–2474.

Dumont, H. J. (1998). The Caspian Lake: history, biota, structure, and function. *Limnol Oceanogr, 43*, 44–52.

Elton, C. S. (1958). *The ecology of invasions by animals and plants*. London, UK: Metheun and Co., Ltd., 181 pp.

Feria, T.P. & Peterson, A. T. (2002). Using point occurrence data and inferential algorithms to predict local communities of birds. *Diversity and Distributions, 8,*49–56.

Fitzpatrick, M. C., Weltzin, F. F. & Sanders, N. J. et al. (2007). The biogeography of prediction error: Why does the introduced range of the fire ant over-predict its native range? *Global Ecol and Biogeogr, 16*, 24–33.

Frammandearter. (2008). List of introduced marine species (Sweden): Caspian mud shrimp *(Corophium curvispinum)*. Available online at http://www.frammandearter.se/0/2english/pdf/Corophiumcurvispinum.pdf (accessed April 2008).

French, J. R. P. III. & Jude, D. J. (2001). Diets and diet overlap of nonindigenous gobies and small benthic native fishes co-inhabiting the St. Clair River, Michigan. J. *Great Lakes Res, 27*(3), 300-311.

Froese, R. & Pauly, D. Editors. (2007). FishBase version 11/2007: *Cottus gobio*, Bullhead. Available online at http://www.fishbase.org/Summary/SpeciesSummary.php?id=2439 (accessed December 19, 2007).

Fuller, P. & Jacobs, G. (2008). *Gymnocephalus cernuus*. USGS Nonindigenous Aquatic Species Database, Gainesville, FL. Available online at <http://nas.er.usgs.gov/ queries/FactSheet.asp?speciesID =7> (revision date: 7/2/2007).

GLBWWG (Great Lakes Ballast Water Working Group). (2008). 2007 summary of Great Lakes ballast water management. Available on-line at http://www.greatlakes-seaway.com/en/pdf/Summ_Ballast_Exams_Report_2007.pdf (accessed July 23, 2008).

Grigorovich, I. A., Colautti, R. I. & Mills, E. L. et al. (2003a). Ballast-mediated animal introductions in the Laurentian Great Lakes: retrospective and prospective analyses. *Can J Fish Aquat Sci, 60*, 740–756.

Grigorovich, I. A., Korniushin, A. V. & Gray, D. K. et al. (2003b). Lake Superior: an invasion coldspot? *Hydrobiol, 499*(1), 191–210.

Hayes, K. R. & Hewitt, C. L. (2000). *Risk assessment framework for ballast water introductions* – Volume II. CRIMP Technical Report 21, Centre for Research on Introduced Marine Pests, CSIRO Marine Research, Hobart, Australia.

Heikkinen, R. K., Luoto, M. & Araujo, M. B. et al. (2006). Methods and uncertainties in bioclimatic envelope modelling under climate change. *Prog Phys Geogr, 30,* 751–777.

Herborg, L.M., Jerde, C. L. & Lodge, D. M. et al. (2007). Predicting invasion risk using measures of introduction effort risk and environmental niche models. *Ecol Appl, 17*(3), 663–674.

Holeck, K. T., Mills, E. L., MacIsaac & H. J. et al. (2004). Bridging troubled waters: biological invasions, transoceanic shipping, and the Laurentian Great Lakes. *Biol Sci, 54*(10), 919–929.

Holeck, K. T., Mills, E. L. & MacIsaac, H. J. et al. (2005). When did the discovery rate for invasive species in the North American Great Lakes accelerate? *Biol Sci, 55*, 4–5.

Holland, J. (1975). *Adaptation in natural and artificial systems*. Ann Arbor, MI: The University of Michigan Press.

Holland, J. (1992). Genetic algorithms. *Scientific American, 267*(1), 66–72.

IAGLR (International Association for Great Lakes Research). (2002). Research and management priorities for aquatic invasive species in the Great Lakes. Available online at http://iaglr.org/scipolicy/ais/index.php (Revision date November 15, 2002).

IPCC (Intergovernmental panel on climate change). (2007). Climate change 2007 – Impacts, adaptation and vulnerability. Contribution of Working Group II to the Fourth Assessment Report of the I PCC.

IUCN (International Union for Conservation of Nature). (2006). Global invasive species database: *Cercopagis pengoi* (crustacean). Available online at http://www.issg.org/database/species/ecology (accessed February 2008).

Johengen, T., Reid, D. & Fahnenstiel, G. et al. (2005). *Assessment of transoceanic NOBOB vessels and low-salinity ballast water as vectors for non-indigenous species introductions to the Great Lakes NOAA*. Great Lakes Environmental Research Laboratory and University of Michigan, Cooperative Institute for Limnology and Ecosystems Research, Ann Arbor, pp. 287.

Jude, D. J., Reider, R. H. & Smith, G. R. (1992). Establishment of Gobiidae in the Great Lakes basin. *Can J Fish Aquat Sci, 49*, 416–421.

Kerr, S. J., Brousseau, C. S. & Muschett, M. (2005). Invasive aquatic species in Ontario: a review and analysis of potential pathways. *Fish, 30*, 21–30.

Kluza, D. A., Vieglais, D. A. & Andreasen, J. K. et al. (2007). Sudden oak death: geographic risk estimates and predictions of origins. *Plant Pathol, 56,* 580–587.

Kolar, C. S. & Lodge, D. M. (2001). Progress in invasion biology: predicting invaders. *TRENDS in Ecol Evol, 16*(4), 199–204.

Leigh, P. (1998). Benefits and costs of the ruffe control program for the Great Lakes fishery. *J Great Lakes Res, 24*(2), 351-360.

Leppakoski, E., Gollasch, S. & Gruszka, P. et al. (2002). The Baltic – a sea of invaders. *Can J Fish Aquat Sci, 59*(7),1175–1188.

Leung, B. & Mandrak, N. E. (2007). The risk of establishment of aquatic invasive species: joining invisibility and propagule pressure. *Proc R Soc Bull, 274*, 2603–2609.

Levri, E. P., Kelly, A. A. & Love, E. (2007). The invasive New Zealand Mud Snail (*Potamopyrgus antipodarum*) in Lake Erie. *JGreat Lakes Res, 33*,1–6.

Lockwood, J. L., Cassey, P. & Blackburn, T. (2005). The role of propagule pressure in explaining species invasions. *TRENDS in Ecol Evol 20(5):*223–228.

MacIsaac, H. J. (1994). Comparative growth and survival of *Dreissena polymorpha* and *Dreissena bugensis*, exotic mollusks introduced to the Great Lakes. *J Great Lakes Res, 20*(4), 783–790.

MacIsaac, H. J., Colautti, R. I. & Niimi, A. J. et al. (2001). Spatial and temporal analysis of transoceanic shipping vectors to the Great Lakes. *Proceedings of the Second International Conference on Marine Bioinvasions*, April 9–11, New Orleans, LA.

MacIsaac, H. J., Robbins, T. C & Lewis, M. A. (2002). Modeling ships' ballast water as invasion threats to the Great Lakes. *Can J Fish Aquat Sci, 59*, 1245–1256.

Mack, R. N. (1996). Predicting the identity and fate of plant invaders: emergent and emerging approaches. *Biol Conserv, 78*(1-2), 107–121.

Marchetti, M. P., Light, T. & Moyle, P. B. et al. (2004). Fish invasions in California watersheds: testing hypotheses using landscape patterns. *Ecol Appl, 14*,1507–1525.

Marsden, J. E. & Jude, D. J. (1995). *Round gobies invade North America*. Fact sheet produced by Sea Grant at Ohio State University, Columbus, OH.

Melina, E. & Rasmussen, J. B. (1994). Patterns in the distribution and abundance of zebra mussel (*Dreissena polymorpha*) in rivers and lakes in relation to substrate and other physicochemical factors. *Can J Fish Aquat Sci, 51*, 1024–1036.

Menges, E. S. (1998). Evaluating extinction risks in plant populations. In: Fielder P.L; Karieva, PM; ed. Conservation biology. New York, NY: Chapman and Hall; pp 49–65.

Menges, E. S. (2000). Population viability analyses in plants: challenges and opportunities. *Trends in Ecology and Evolution, 15*, 51–56.

MDEQ (Michigan Department of Environmental Quality). (2008). Ballast water control general permit. Available online at http://www.michigan.gov/deq/0,1607,7-135-3313_3682_3713- 153446--,00.html (accessed August 7, 2008).

Mills, E. L., Leach, J H. & Carlton, J. T. et al. (1993). Exotic species in the Great Lakes: a history of biotic crises and anthropogenic introductions. *J Great Lakes Res, 19*(1),1–54.

Mills, E L., Leach, J. H. & Carlton, J. T. et al. (1994). Exotic species and the integrity of the Great Lakes: lessons from the past. *Biol Sci, 44*(10), 666–676.

Mills, E. L., Rosenberg, G. & Spidle, A. P. et al. (1996). A review of the biology and ecology of the quagga mussel (*Dreissena bugensis*), a second species of freshwater Dreissenid introduced to North America. *Amer Zool, 36*, 271–286.

Mills, E. L., Chrisman, J. R. & Baldwin, B. et al. (1999). Changes in the Dressenid community in the lower Great Lakes with emphasis on Southern Lake Ontario. *J Great Lakes Res, 25*(1), 187–197.

National Invasive Species Council. (2007). 2008–2012 National invasive species management plan, draft for public comment, December 20, 2007. Available online at http://www.invasivespeciesinfo.gov/council/mpdraft07.pdf (accessed April 2008).

NBIC (National Ballast Information Clearinghouse). (2008). NBIC Online Database. Electronic publication, Smithsonian Environmental Research Center & United States Coast Guard. Available online at http://invasions.si.edu/nbic/search. (accessed July 2008).

Nico, L. Fuller, P. & Jacobs, G. (2008). *Scardinius erythrophthalmus*. USGS Nonindigenous Aquatic Species Database, Gainesville, FL. Available online at http://nas.er.usgs.gov/ queries/FactSheet.asp?speciesID=648 (revision date: 8/21/2007).

Nico, L. & Fuller, P. (2008). *Tinca tinca*. USGS Nonindigenous Aquatic Species Database, Gainesville, FL. Available online at http://nas.er.usgs.gov/queries/FactSheet.asp ?speciesID=652 (revision date: 4/24/2006).

NOAA (National Oceanic Atmospheric Administration). (2007a). Great Lakes list. National Center for Research on Aquatic Invasive Species. Available online at http://www.glerl.noaa.gov/res/Programs/ncrais/docs/great-lakes (accessed March 5, 2008).

NOAA (National Oceanic Atmospheric Administration). (2007b). Sea Grant National: Ballast water technology projects funded by NOAA and FWS. Available online at http://www.seagrant.noaa.gov/roe/ballastwatertechprojects.html (accessed August 15, 2007).

NOAA (National Oceanic and Atmospheric Administration). (2008). Great Lakes bathymetry. National Environmental Satellite, Data, and Information Service. Available online at http://map.ngdc.noaa.gov/website/mgg/greatlakesbathy/viewer.htm (accessed February 2008).

NRC (National Research Council). (2008). Great Lakes shipping, trade, and aquatic invasive species, Transportation Research Board, Special Report 291. Available online at http://onlinepubs.trb.org/Onlinepubs/sr/sr291prepub.pdf (accessed July 23, 2008).

O'Neill, C. R., Jr. (2004). Biological pollution of the Great Lakes: the nonindigenous aquatic species invasion of North America's freshwater seas. New York Sea Grant. Available online at http://www.sgnis.org/publicat/papers/neil2004.pdf (accessed March 2008).

Owens, R. W., O'gorman, R. & Mills, E. L. et al. (1998). Blueback herring (*Alosa aestivalis*) in Lake Ontario: First record, entry route, and colonization potential. *J Great Lakes Res, 24*(3), 723–730.

Peterson, A. T. (2003). Predicting the geography of species' invasions via ecological niche modeling. *Q Rev Biol, 78*(4), 419–433.

Peterson, A. T. & Cohoon, K. P. (1999). Sensitivity of distributional prediction algorithms to geographic data completeness. *Ecol Model, 117,*159–164.

Peterson, A. T. & Vieglais, D. A. (2001). Predicting species invasions using ecological niche modeling: new approaches from bioinformatics attack a pressing problem. *Biol Sci, 51*(5), 363– 371.

Peterson, A. T., Soberon, J. & Sanchez-Cordero, V. (1999). Conservatism of ecological niches in evolutionary time. *Science, 285,* 1265–1267.

Peterson, A. T., Stockwell, D. R. B. & Kluza, D. A. (2002). Distributional prediction based on ecological niche modeling of primary occurrence data. In: Scott, JM; Heglund, PJ; Morrison, ML; eds. Predicting species occurrences: issues of scale and accuracy. *Washington,* DC: Island Press, pp. 617–623.

Pimentel, D., Lach, L. & Zuniga, R. et al. (2000). Environmental and economic costs of nonindigenous species in the United States. *Biol Sci, 50*, 53–65.

Pimentel, D. (2005). Aquatic nuisance species in the New York State Canal and Hudson River Systems and the Great Lakes Basin: an economic and environmental assessment. *Environ Manage, 35*(5), 692–701.

Ramcharan, C. W., Padilla, D. K. & Dodson, S. I. (1992). Models to predict potential occurrence and density of the zebra mussel, *Dreissena polymorpha. Can J Fish Aquat Sci, 49,* 2611–2620.

Rasmussen, J. L. (2002). The Cal-Sag and Chicago sanitary and ship canal: a perspective on the spread and control of selected aquatic nuisance fish species. U.S. Fish and Wildlife Service, La Crosse, Fisheries Resource Office. Available online at http://www.fws.gov/midwest/LaCrosseFisheries/reports/Connecting_Channels_Paper_Final.pdf (accessed March 2008).

Reid, D. F. & Orlova, M. I. (2002). Geological and evolutionary underpinnings for the success of Ponto-Caspian species invasions in the Baltic Sea and North American Great Lakes. *Can J Fish Aquat Sci, 59*, 1144–1158.

Ricciardi, A. (2006). Patterns of invasion of the Laurentian Great Lakes in relation to changes in vector activity. *Divers Distrib, 12*, 425–433.

Ricciardi, A. & MacIsaac, H. J. (2000). Recent mass invasion of the North American Great Lakes by Ponto-Caspian species. *TREE, 15*(2), 62–65.

Ruiz, G. & Reid, D. F. (2007). Current state of understanding about the effectiveness of ballast water exchange (BWE) in reducing aquatic nonindigenous species (ANS) introductions to the Great Lakes Basin and Chesapeake Bay, USA: synthesis and analysis of existing information.

National Oceanic and Atmospheric Administration, Technical Memorandum GLERL-142, Great Lakes Environmental Research Lab, Ann Arbor, Michigan. 127 pp.

Sakai, A. K., Allendorf, F. W. & Holt, J. S. et al. (2001). The population biology of invasive species. *Ann Rev Ecol and Syst, 32*, 305–332.

Sala, O. E., Chapin, F. S. III. & Armesto, J. et al. (2000). Global biodiversity scenarios for the year 2100. *Science, 287*, 1770–1774.

Sandlund, O. T., Naesje, T. F. & Klyke, L. et al. (1985). The vertical distribution of fish species in Lake Mjosa, Norway as shown by gillnet catches and echo sounder. *Inst Freshwater Res – Drottningholm, 62,*136–149.

Scachetti-Pereira, R. (2002). Desktop GARP version 1.1.6. Available online at http://www.lifemapper.org/desktopgarp/ (accessed April 3, 2006).

Simberloff, D. & Von Holle, B. (1999). Positive interactions of nonindigenous species: invasional meltdown? *Biol Inv, 1*, 21–32.

Sing, T., Sander, O. & Beerenwinkel N. et al. (2005). ROCR: visualizing classifier performance in R. *Bioinformatics, 21,* 3940–3941.

Smith, L. D., Wonham, M. J. & McCann, L. D. et al. (1999). Invasion pressure to a ballast-flooded estuary and an assessment of inoculant survival. *Biol Inv 1*:67–87.

Stockwell, D. R. B. & Peters, D. (1999). The GARP modeling system: problems and solutions to automated spatial prediction. *Int J Geogr Inf Syst, 13*(2),143–158.

Stockwell, D. R. B., & Peterson, A. T. (2002). Effects of sample size on accuracy of species distribution models. *Ecol Model, 148*(1),1–13.

Sutherst, R. W., Maywald, G. F. & Yonow, T. et al. (1999). CLIMEX user guide—predicting the effects of climate on plants and animals. Collingwood, Victoria, Australia: CSIRO Publishing.

Swets, J. A. (1988). Measuring the accuracy of diagnostic systems. *Science 240:*1285–1293.

Theoharides, K. A. & Dukes, J. S. (2007). Plant invasion across space and time: factors affecting nonindigenous species success during four stages of invasion. *New Phytol, 176,* 256–273.

Torchin, M. E. & Mitchell, C. E. (2004). Parasites, pathogens, and invasions by plants and animals. *Front Ecol Env, 2*,183–190.

University of Minnesota. (2006). *Cercopagis pengoi* invades Lake Ontario fact sheet: October 1999. Ontario Federation of Anglers and Hunters, University of Minnesota. Available online at http://www.seagrant.umn.edu/exotics/fishhook.html (accessed February 2008).

USCG (United States Coast Guard). (2001). Report to Congress on the voluntary national guidelines for ballast water management. United States Coast Guard Office of Operating and Environmental Standards. Available online at http://www.USCG.mil/hq/g-m/mso/bwm.htm (accessed May 1, 2007).

U.S. Congress. (1993). Harmful nonindigenous species in the United States. Office of Technology Assessment, Superintendent of Documents, P.O. Box 371954, Pittsburgh, PA 15250-7974; OTA-F-565 (phone 202-512-1800).

USGS (United States Geological Survey). (2007). Nonindigenous aquatic species database. Maps, Imagery, and Publication. Available online at http://www.usgs.gov/pubprod/maps.html (accessed 12/04/07).

USGS (United States Geological Survey). (2008a). Great Lakes Science Center: Round Goby. Available online at http://www.glsc.usgs.gov/main.php?content=research_invasive_goby&title=Invasive%20Fish0& menu=research invasive fish (revision date: January 31, 2008).

USGS (United States Geological Survey). (2008b). Nonindigenous aquatic species: Cercopagis pengoi. Available online at http://nas.er.usgs.gov/queries/FactSheet.asp?speciesID=163 (revision date: April 25, 2008).

USGS (United States Geological Survey). (2008c). NAS nonindigenous aquatic species: Dreissena polymorpha. Available online at http://nas.er.usgs.gov/queries/FactSheet.asp?speciesID=5 (Revision Date: 1/18/2008).

Vanderploeg, H. A., Liebig, J. R. & Carmichael, W. W. et al. (2001). Zebra mussel (Dreissena polymorpha) selective filtration promoted toxic microcysts blooms in Saginaw Bay (Lake Huron) and Lake Erie. *Can J Fish Aquat Sci, 58*, 1208–1221.

VA Inland Fisheries (Virginia Department of Inland Fisheries). (2008). Blueback herring. Available online at http://www.dgif.state.va.us/wildlife (accessed February 2008).

Watkins, J. M., Dermott, R. & Lozano, S. J. et al. (2007). Evidence for remote effects of Dreissenid mussels on the amphipod *Diporeia*: Analysis of Lake Ontario benthic surveys 1972–2003. *J Great Lakes Res, 33*, 642–657.

Wilcove, D. S., Rothstein, D. & Dubow J. et al. (1998). Quantifying threats to imperiled species in the United States. *Biol Sci, 48*(8), 607–615.

Williamson, M. (1996). Biological invasions. Chapman and Hall: London.

Wisconsin, D. N. R. (Wisconsin Department of Natural Resources). (2008). Quagga mussel. Available on-line at http://www.dnr.state.wi.us/org/caer/ce/eek/critter/invert/quaggamussel.htm (accessed February 2008).

Wisconsin Sea Grant. (2008). Round Goby, fish of the Great Lakes. University of Wisconsin Sea Grant Institute. Available online at http://www.seagrant.wisc.edu/greatlakesfish/roundgoby.html_ (accessed March 2008).

Wonham, M. J., Walton, W. C. & Ruiz, G. M. et al. (2001). Going to the source: role of the invasion pathway in determining potential invaders. *Marine Ecology Progress Series, 215*, 1–12.

WSR (World Shipping Register). (2007). World shipping register: world sea ports. Available online at http://www.world-register.com/ports.php (accessed February 2, 2007).

Zaranko, D. T., Farara, D. G. & Thompson, F. G. (1997). Another exotic mollusk in the Laurentian Great Lakes: the New Zealand native *Potamopyrgus antipodarum* (Gray 1843) (Gastropoda, Hydrobiidae). *Canadian Journal of Fisheries and Aquatic Sciences, 54,* 809–814.

Appendix A. List of Nonindigenous Species That Have Been Reported as Occurring in the Great Lakes

Year of Invasion	Species	Common Name	Type	Endemic Region	Location of First Sighting	Vector
1840	*Rumex obtusifolius*	bitter dock	Plant	Eurasia	Widespread	Unknown
1843	*Echinochloa crusgalli*	barnyard grass	Plant	Eurasia	Widespread	Release (deliberate)
1843	*Solanum dulcamara*	bittersweet nightshade	Plant	Eurasia	Widespread	Release (deliberate)
1843	*Mentha piperita*	peppermint	Plant	Eurasia	Widespread	Release (deliberate)
1843	*Conium maculatum*	poison hemlock	Plant	Eurasia	Widespread	Release (deliberate)
1843	*Poa trivalis*	rough-stalked meadow grass	Plant	Eurasia	Widespread	Release (deliberate)
1843	*Mentha spicata*	spearmint	Plant	Eurasia	Widespread	Release (deliberate)
1843	*Polygonum persicaria*	lady's thumb	Plant	Eurasia	Widespread	Unknown
1847	*Rorippa nasturtium-a quaticum*	water cress	Plant	Eurasia	Lake Ontario	Release (deliberate)
1860	*Elimia virginica*	snail	Mollusk	Atlantic NA	Erie Canal	Canals
1862	*Juncus gerardii*	black-grass rush	Plant	Atlantic NA	Chicago	Shipping, Solid Ballast
1864	*Najas marina*	spiny naiad	Plant	Eurasia	Lake Ontario drainage	Shipping, Solid Ballast
1865	*Sonchus arvensis*	field sow thistle	Plant	Eurasia	Central NY	Release (unintentional)
1866	*Carex disticha*	sedge	Plant	Eurasia	Lake Ontario	Shipping, Solid Ballast
1867	*Chenopodium glaucum*	oak leaved goose foot	Plant	Eurasia	Lake Ontario drainage	Railroads and Highways
1869	*Lythrum salicaria*	purple loosestrife	Plant	Eurasia	Lake Ontario	Canals
1871	*Bithynia tentaculata*	faucet snail	Mollusk	Eurasia	Lake Michigan	Shipping, Solid Ballast
1873	*Alosa pseudoharengus*	alewife	Fish	Atlantic NA	Lake Ontario	Canals

(Continued)

Year of Invasion	Species	Common Name	Type	Endemic Region	Location of First Sighting	Vector
1873	*Oncorhynchus tshawytscha*	chinook salmon	Fish	Pacific NA	Widespread	Release (deliberate)
1874	*Epilobium hirsutum*	great hairy willow herb	Plant	Eurasia	Lake Ontario	Release (unintentional)
1876	*Oncorhynchus mykiss*	rainbow trout	Fish	Pacific NA	Lake Huron drainage	Release (deliberate)
1878	*Carassius auratus*	goldfish	Fish	Asia	Widespread	Release (unintentional)
1879	*Cyprinus carpio*	common carp	Fish	Ponto-Caspian	Widespread	Release (deliberate)
1879	*Potamogeton crispus*	curlyleaf pondweed	Plant	Eurasia	Lake Ontario drainage	Release (deliberate)
1880	*Typha angustifolia*	narrow leaved cattail	Plant	Eurasia	Lake Ontario	Canals
1882	*Lysimachia nummularia*	moneywort	Plant	Eurasia	Lake Ontario	Release (deliberate)
1882	*Alopecurus geniculatus*	water foxtail	Plant	Eurasia	Lake Erie	Release (deliberate)
1883	*Salmo trutta*	brown trout	Fish	Eurasia	Lake Michigan drainage	Release (deliberate)
1884	*Agrostis gigantea*	redtop	Plant	Eurasia	Lake Superior	Release (deliberate)
1884	*Rorippa sylvestris*	creeping yellow cress	Plant	Eurasia	Lake Ontario	Shipping, Solid Ballast
1886	*Salix fra gilis*	crack willow	Plant	Eurasia	Widespread	Release (deliberate)
1886	*Salix purpurea*	purple willow	Plant	Eurasia	Widespread	Release (deliberate)
1886	*Myosotis scorpioides*	true forgot-me-not	Plant	Eurasia	Lake Ontario	Release (deliberate)
1886	*Salix alba*	white willow	Plant	Eurasia	Widespread	Release (deliberate)
1886	*Iris pseudacorus*	yellow flag	Plant	Eurasia	Lake Ontario	Release (deliberate)
1892	*Lycopus asper*	western water horehound	Plant	Mississippi R.	Lake Erie	Release,(unintention al)
1893	*Puccinellia distans*	weeping alkali grass	Plant	Eurasia	Lake Ontario	Shipping, Solid Ballast

(Continued)

Year of Invasion	Species	Common Name	Type	Endemic Region	Location of First Sighting	Vector
1894	*Stellaria aquatica*	giant chickweed	Plant	Eurasia	Lake St. Clair	Unknown
1895	*Juncus compressus*	flattened rush	Plant	Eurasia	Lake Ontario drainage	Release (unintentional)
1895	*Pisidium moitessierianum*	pea clam	Mollusk	Eurasia	Lake Superior	Shipping, Solid Ballast
1896	*Carex flacca*	sedge	Plant	Eurasia	Detroit River	Unknown
1897	*Valvata piscinalis*	European valve snail	Mollusk	Eurasia	Lake Ontario	Shipping, Solid Ballast
1897	*Pisidium amnicum*	pea clam	Mollusk	Eurasia	Lake Ontario	Shipping, Solid Ballast
1901	*Rumex longifolius*	yard dock	Plant	Eurasia	Lake Superior	Release (deliberate)
1901	*Radix auricularia*	European ear snail	Mollusk	Eurasia	Lake Michigan	Release (unintentional)
1902	*Aeromonas salmonicida*	furunculosis	Other Invertebrate	Unknown	Unknown	Release (unintentional)
1902	*Sonchus arvensis* var. *glabrescens*	smooth field sow thistle	Plant	Eurasia	Lake Erie	Release (unintentional)
1903	*Lycopus europaeus*	European water horehound	Plant	Eurasia	Lake Ontario	Shipping, Solid Ballast
1905	*Butomus umbellatus*	flowering rush	Plant	Eurasia	Detroit River	Shipping, Solid Ballast
1906	*Viviparus georgianus*	banded mystery snail	Mollusk	Mississippi R.	Lake Michigan	Aquarium release
1912	*Impatiens glandulifera*	Indian balsam	Plant	Asia	Lake Huron	Release (deliberate)
1912	*Osmerus mordax*	rainbow smelt	Fish	Atlantic NA	Lake Michigan drainage	Release (deliberate)
1913	*Alnus glutinosa*	black alder	Plant	Eurasia	Widespread	Release (deliberate)
1913	*Lysimachia vulgaris*	garden loosetrife	Plant	Eurasia	Lake Ontario	Release (deliberate)
1913	*Rhamnus frangula*	glossy buckthorn	Plant	Eurasia	Lower Great Lakes, Ontario	Release (deliberate)
1915	*Mentha gentilis*	creeping whorled mint	Plant	Eurasia	Lake Ontario	Release (deliberate)

(Continued)

Year of Invasion	Species	Common Name	Type	Endemic Region	Location of First Sighting	Vector
1915	*Veronica beccabunga*	European brookline	Plant	Eurasia	Lake Ontario	Shipping, Solid Ballast
1916	*Pluchea odorata* var. *purpurescens*	salt-marsh fleabane	Plant	Atlantic NA	Lake Erie drainage	Release (unintentional)
1916	*Pisidium henslowanum*	henslow's pea clam	Mollusk	Eurasia	Lake Ontario	Shipping, Solid Ballast
1918	*Gillia altilis*	snail	Mollusk	Atlantic NA	Lake Ontario drainage	Canals
1922	*Juncas inflexus*	rush	Plant	Eurasia	Lake Ontario drainage	Unknown
1923	*Gambusia affinis*	western mosquitofish	Fish	Mississippi R.	Lake Michigan drainage	Release (deliberate)
1924	*Sphaerium corneum*	fingernail clam	Mollusk	Eurasia	Lake Ontario drainage	Shipping, Ballast Water
1925	*Marsilea quadrifolia*	European water clover	Plant	Eurasia	Lake Ontario drainage	Release (deliberate)
1926	*Enteromorpha intestinalis*	green alga	Benthic Alga	Atlantic NA	Lake Ontario	Release (unintentional)
1927	*Acentropus niveus*	aquatic moth	Other Invertebrate	Eurasia	St. Lawrence R., Montreal	Release (unintentional)
1928	*Noturus insignis*	margined madtom	Fish	Atlantic NA	Lake Ontario drainage	Canals
1928	*Lepomis microlophus*	redear sunfish	Fish	Southern U.S.	Lake Michigan drainage	Release (deliberate)
1929	*Lepomis humilis*	orange spotted sunfish	Fish	Mississippi R.	Lake Erie drainage	Canals

(Continued)

Year of Invasion	Species	Common Name	Type	Endemic Region	Location of First Sighting	Vector
1930	*Nymphoides peltata*	yellow floating heart	Plant	Eurasia	Lake Erie drainage	Release (unintentional)
1931	*Cipangopaludina chinensis malleata*	Oriental mystery snail	Mollusk	Asia	Niagara River	Release (aquarium)
1932	*Najas minor*	minor naiad	Plant	Eurasia	Lake Erie drainage	Release (deliberate)
1933	*Oncorhynchus kisutch*	coho salmon	Fish	Pacific	Lake Erie	Release (deliberate)
1933	*Craspedacusta sowerbyi*	freshwater jellyfish	Other Invertebrate	Asia	Lake Erie drainage	Release (unintentional)
1934	*Lophopodella carteri*	bryozoan	Other Invertebrate	Asia	Lake Erie	Canals
1935	*Cabomba caroliniana*	fanwort	Plant	Southern U.S.	Lake Michigan drainage	Release (aquarium)
1936	*Sparganium glomeratum*	bur reed	Plant	Eurasia	Lake Superior	Unknown
1938	*Actinocyclus normanii* fo. *subsalsa*	diatom	Phytoplankton	Eurasia	Lake Ontario	Shipping, Ballast Water
1938	*Diatoma ehrenbergii*	diatom	Phytoplankton	Widespread	Lake Michigan	Shipping, Ballast Water
1938	*Stephanodiscus binderanus*	diatom	Phytoplankton	Eurasia	Lake Michigan	Shipping, Ballast Water
1939	*Misgurnus anguillicaudatus*	Oriental weatherfish	Fish	Asia	Lake Huron drainage	Release (unintentional)
1940	*Cipangopaludina japonica*	Oriental mystery snail	Mollusk	Asia	Lake Erie	Release (deliberate)
1940	*Glyceria maxima*	reed sweet-grass	Plant	Eurasia	Lake Ontario	Release (deliberate)
1943	*Tanysphyrus lemnae*	aquatic weevil	Other Invertebrate	Eurasia	Unknown	Unknown

(Continued)

Year of Invasion	Species	Common Name	Type	Endemic Region	Location of First Sighting	Vector
1946	*Cyclotella pseudostelligera*	diatom	Phytoplankton	Widespread	Lake Michigan	Shipping, Ballast Water
1946	*Stephanodiscus subtilis*	diatom	Phytoplankton	Eurasia	Lake Michigan	Shipping, Ballast Water
1950	*Phenacobius mirabilis*	suckermouth minnow	Fish	Mississippi R.	Lake Erie drainage	Canals
1950	*Morone americana*	white perch	Fish	Atlantic NA	Lake Ontario	Canals
1950	*Oncorhynchus nerka*	kokanee	Fish	Pacific NA	Lake Ontario drainage	Release (deliberate)
1950	*Cirsium palustre*	marsh thistle	Plant	Eurasia	Lake Superior	Unknown
1950	*Potamothrix bedoti*	oligochaete	Annelid	Ponto-Caspian	Lake Ontario	Unknown
1950	*Pluchea odorata* var. *succulenta*	salt-marsh fleabane	Plant	Atlantic NA	Lake Ontario drainage	Unknown
1951	*Branchiura sowerbyi*	oligochaete	Annelid	Asia	Lake Michigan drainage	Release (unintentional)
1951	*Carex acutiformis*	swamp sedge	Plant	Eurasia	Lake Michigan drainage	Unknown
1952	*Myriophyllum spicatum*	Eurasian watermilfoil	Plant	Eurasia	Lake Erie	Release (aquarium)
1952	*Potamothrix moldaviensis*	oligochaete	Annelid	Ponto-Caspian	Lake Ontario	Unknown
1956	*Cordylophora caspia*	hydroid	Other Invertebrate	Ponto-Caspian	Lake Erie	Release (unintentional)
1956	*Oncorhynchus gorbuscha*	pink salmon	Fish	Pacific NA	Lake Superior	Release (unintentional)
1958	*Eurytemora affinis*	calanoid copepod	Zooplankton (crustacean)	Widespread	Lake Ontario	Shipping, Ballast Water

(Continued)

Year of Invasion	Species	Common Name	Type	Endemic Region	Location of First Sighting	Vector
1959	*Trapa natans*	water chestnut	Plant	Eurasia	Lake Ontario drainage	Release (aquarium)
1959	*Lasmigona subviridis*	mussel	Mollusk	Atlantic NA	Lake Ontario drainage	Canals
1959	*Pisidium supinum*	humpback pea clam	Mollusk	Europe	Lake Ontario	Shipping, Ballast Water
1960	*Glugea hertwigi*	protozoan	Other Invertebrate	Eurasia	Lake Erie	Release (unintentional)
1960	*Polygonum caespitosum* var. *lon gisetum*	Bristly Lady's Thumb	Plant	E. Asia	Lake Erie drainage	Unknown
1962	*Lepisosteus platostomus*	shortnose gar	Fish	Mississippi R.	Lake Michigan drainage	Canals
1962	*Thalassiosira weissflogii*	diatom	Phytoplankton	Widespread	Detroit River	Shipping, Ballast Water
1963	*Skeletonema potamos*	diatom	Phytoplankton	Widespread	Lake Erie drainage	Shipping, Ballast Water
1964	*Bangia atropurpurea*	red alga	Benthic Alga	Atlantic NA	Lake Erie	Shipping
1964	*Cyclotella atomus*	diatom	Phytoplankton	Widespread	Lake Michigan	Shipping, Ballast Water
1964	*Cyclotella cryptica*	diatom	Phytoplankton	Widespread	Lake Michigan	Shipping, Ballast Water
1964	*Cyclotella woltereki*	diatom	Phytoplankton	Widespread	Lake Michigan	Shipping, Ballast Water
1964	*Chroodactylon ramosum*	red alga	Benthic Alga	Atlantic Ocean	Lake Erie	Shipping, Ballast Water

(Continued)

Year of Invasion	Species	Common Name	Type	Endemic Region	Location of First Sighting	Vector
1965	*Potamothrix vejdovskyi*	oligochaete	Annelid	Ponto-Caspian	Lake Erie	Shipping? Ballast Water?
1966	*Eubosmina coregoni*	waterflea	Zooplankton (crustacean)	Eurasia	Lake Michigan	Shipping, Ballast Water
1966	*Epiobium parviflorum*	small flowered hairy willow herb	Plant	Eurasia	Lake Michigan drainage	Unknown
1967	*Skistodiaptomus pallidus*	calanoid copepod	Zooplankton (crustacean)	Mississippi R.	Lake Ontario	Release (unintentional)
1968	*Myxobolus (Myxosoma) cerebralis*	salmonid whirling disease	Other Invertebrate	Unknown	Lake Erie drainage	Release (unintentional)
1968	*Dugesia polychroa*	flatworm	Other Invertebrate	Europe	Lake Ontario	Shipping, Ballast Water
1969	*Solidago sempervirens*	seaside goldenrod	Plant	Atlantic NA	Lake Michigan	Release (unintentional)
1971	*Enneacanthus gloriosus*	bluespotted sunfish	Fish	Atlantic NA	Lake Ontario drainage	Release (aquarium)
1972	*Cyclops strenuus*	copepod	Zooplankton (crustacean)	Hudson Bay	Lake Superior	Canals (water diversion)
1972	*Hydrocharis morsus-ranae*	European frogbit	Plant	Eurasia	Lake Ontario	Release (unintentional)
1973	*Skeletonema subsalsum*	diatom	Phytoplankton	Eurasia	Lake Erie	Shipping, Ballast Water
1973	*Thalassiosira guillardii*	diatom	Phytoplankton	Widespread	Lake Erie	Shipping, Ballast Water
1973	*Thalassiosira pseudonana*	diatom	Phytoplankton	Widespread	Lake Erie drainage	Shipping, Ballast Water

(Continued)

Year of Invasion	Species	Common Name	Type	Endemic Region	Location of First Sighting	Vector
1973	*Nitocra hibernica*	harpacticoid copepod	Benthic Crustacean	Eurasia	Lake Ontario	Shipping, Ballast Water
1975	*Sphacelaria fluviatilis*	brown alga	Benthic Alga	Asia	Lake Michigan drainage	Release (aquarium)
1975	*Lotus corniculatus*	birdsfoot trefoil	Plant	Eurasia	Lake Superior	Release (deliberate)
1975	*Renibacterium (Corynebacterium) salmoninarum*	bacterial kidney disease	Bacteria	Unknown	Lake Superior	Release (unintentional)
1975	*Sphacelaria lacustris*	brown alga	Benthic Alga	Unknown	Lake Michigan	Shipping, Ballast Water
1975	*Hymenomonas roseola*	cocco-lithophorid alga	Phytoplankton	Eurasia	Lake Huron	Shipping, Ballast Water
1978	*Biddulphia laevis*	diatom	Phytoplankton	Widespread	Lake Michigan	Shipping, Ballast Water
1978	*Chaetoceros hohnii*	diatom	Phytoplankton	Unknown	Lake Huron	Shipping, Ballast Water
1978	*Thalassiosira lacustris*	diatom	Phytoplankton	Eurasia	Lake Erie	Shipping, Ballast Water
1979	*Notropis buchanani*	ghost shiner	Fish	Mississippi R.	Lake St. Clair drainage	Release (bait)
1979	*Enteromorpha prolifera*	green alga	Benthic Alga	Atlantic NA	Lake St. Clair drainage	Unknown
1980	*Corbicula fluminea*	Asiatic clam	Mollusk	East Asia	Lake Erie	Release (aquarium)
1980	*Ripistes parasita*	oligochaete	Annelid	Eurasia	Lake Huron	Shipping, Ballast Water

(Continued)

Year of Invasion	Species	Common Name	Type	Endemic Region	Location of First Sighting	Vector
1980	*Daphnia galeata galeata*	waterflea	Zooplankton (crustacean)	Eurasia	Lake Erie	Shipping, Ballast Water
1982	*Lupinus polyphyllus*	lupine	Plant	Eurasia	Lake Superior	Release (unintentional)
1982	*Bythotrephes lon gimanus*	spiny waterflea	Zooplankton (crustacean)	Eurasia	Lake Ontario	Shipping, Ballast Water
1983	*Nitellopsis obtusa*	green alga	Phytoplankton	Eurasia	Lake St. Clair	Shipping, Ballast Water
1983	*Gianius (Phallodrilus) aquaedulcis*	oligochaete	Annelid	Europe	Niagara River	Shipping, Ballast Water
1985	*Salmin cola lotae*	copepod	Zooplankton (crustacean)	Eurasia	Lake Superior	Unknown
1986	*Gymnocephalus cernuus*	Eurasian ruffe	Fish	Ponto-Caspian	Lake Superior	Shipping, Ballast Water
1986	*Apeltes quadracus*	fourspine stickleback	Fish	Atlantic NA	Lake Superior	Shipping, Ballast Water
1988	*Argulus japonicus*	parasitic copepod	Zooplankton (crustacean)	Asia	Lake Michigan	Release (aquarium)
1988	*Thalassiosira baltica*	diatom	Phytoplankton	Europe	Lake Ontario	Shipping, Ballast Water
1988	*Bosmina maritima*	waterflea	Zooplankton (crustacean)	Eurasia	Lake Erie	Shipping, Ballast Water
1988	*Dreissena polymorpha*	zebra mussel	Mollusk	Ponto-Caspian	Lake St. Clair	Shipping, Ballast Water
1989	*Scardinius erythrophthalmus*	rudd	Fish	Eurasia	Lake Ontario	Release (Bait)

(Continued)

Year of Invasion	Species	Common Name	Type	Endemic Region	Location of First Sighting	Vector
1989	*Dreissena bugensis*	quagga mussel	Mollusk	Ponto-Caspian	Lake Ontario	Shipping, Ballast Water
1990	*Neogobius melanostomus*	round goby	Fish	Ponto-Caspian	St. Clair River	Shipping, Ballast Water
1990	*Proterorhinus marmoratus*	tubenose goby	Fish	Ponto-Caspian	St. Clair River	Shipping, Ballast Water
1991	*Potamopyrgus antipodarum*	New Zealand mud snail	Mollusk	Australasia	Lake Ontario	Shipping, Ballast Water
1992	*Acanthostomum* sp.	digenean fluke	Other Invertebrate	Eurasia	Lake Superior	Shipping, Ballast Water
1992	*Ichthyocotylurus pileatus*	digenean fluke	Other Invertebrate	Ponto-Caspian	Lake Superior	Shipping, Ballast Water
1992	*Neascus brevicaudatus*	digenean fluke	Other Invertebrate	Eurasia	Lake Superior	Shipping, Ballast Water
1992	*Trypanosoma acerinae*	flagellate	Other Invertebrate	Ponto-Caspian	Lake Superior	Shipping, Ballast Water
1992	*Dactylogyrus amphibothrium*	monogenetic fluke	Other Invertebrate	Eurasia	Lake Superior	Shipping, Ballast Water
1992	*Dactylogyrus hemiamphibothrium*	monogenetic fluke	Other Invertebrate	Eurasia	Lake Superior	Shipping, Ballast Water
1994	*Echinogammarus ischnus*	amphipod	Benthic Crustacean	Ponto-Caspian	Detroit River /Lake Erie	Shipping, Ballast Water
1994	*Scolex pleuronectis*	cestode	Other Invertebrate	Ponto-Caspian	Lake St. Clair	Shipping, Ballast Water
1994	*Sphaeromyxa sevastopoli*	mixosporidian	Other Invertebrate	Ponto-Caspian	Lake St. Clair	Shipping, Ballast Water

(Continued)

Year of Invasion	Species	Common Name	Type	Endemic Region	Location of First Sighting	Vector
1994	*Neoergasilus japonicus*	copepod	Zooplankton (crustacean)	Eastern Asia	St. Clair River	Unknown
1994	*Megacyclops viridis*	cyclopoid copepod	Zooplankton (crustacean)	Europe	Lake Huron	Unknown
1995	*Alosa aestivalis*	blueback herring	Fish	Atlantic NA	Lake Ontario	Canals
1996	*Heteropsyllus* nr. *nunni*	harpacticoid copepod	Benthic Crustacean	Atlantic NA?	Lake Michigan	Shipping, Ballast Water
1997	*Acineta nitocrae*	suctorian	Other Invertebrate	Eurasia	Lake Erie	Shipping, Ballast Water
1998	*Cercopagis pengoi*	fish-hook waterflea	Zooplankton (crustacean)	Ponto-Caspian	Lake Ontario	Shipping, Ballast Water
1998	*Schizopera borutzkyi*	harpacticoid copepod	Benthic Crustacean	Ponto-Caspian	Lake Michigan	Shipping, Ballast Water
1999	*Daphnia lumholtzi*	waterflea	Zooplankton (crustacean)	Africa, Australasia	Lake Erie	Release (unintentional)
1999	*Nitocra incerta*	harpacticoid copepod	Benthic Crustacean	Ponto-Caspian	Detroit River	Shipping, Ballast Water
2000	*Heterosporis* sp.	microsporidian	Other Invertebrate	Unknown	Lake Ontario	Unknown
2001	*Rhabdovirus carpio*	spring viraemia of carp (SVC)	Virus	Eurasia	Lake Michigan drainage	Release (aquarium)
2001	*Gammarus tigrinus*	Amphipod	Benthic Crustacean	Atlantic NA	Lake Superior (Lake Huron in 2002)	Shipping, Ballast Water
2001	*Psammonobiotus communis*	testate amoeba	Other Invertebrate	Ponto-Caspian	Lake Ontario	Shipping, Ballast Water

(Continued)

Year of Invasion	Species	Common Name	Type	Endemic Region	Location of First Sighting	Vector
2002	*Ranavirus* sp.	largemouth bass virus	Virus	Unknown	Lake Michigan drainage	Release (unintentional)
2002	*Psammonobiotus linearis*	testate amoeba	Other Invertebrate	Ponto-Caspian	Lake Ontario	Shipping, Ballast Water
2002	*Psammonobiotus dziwnowi*	testate amoeba	Other Invertebrate	PontoCaspian?	Lake Ontario	Shipping, Ballast Water
2002	*Cylindrospermopsis raciborskii*	cyanobacterium	Phytoplankton	South America?	Lake Michigan drainage	Unknown
2002	*Piscirickettsia cf. salmonis*	muskie pox	Other Invertebrate	Unknown	Lake St. Clair	Unknown
2003	*Enteromorpha flexuosa*	green alga	Benthic Alga	Widespread	Lake Michigan drainage	Shipping
2005	*Novirhabdovirus* sp.	VHS	Virus	Atlantic NA?	Lake Ontario	Shipping?, Ballast Water?
2006	*Hemimysis anomala*	bloody-red mysid	Benthic Crustacean	Ponto-Caspian	Lake Michigan drainage	Shipping, Ballast Water

Source: http://www.glerl.noaa.gov/res/Programs/ncrais/docs/great-lakes-list.xls (accessed March 5, 2008).

Appendix B. Nonindigenous Species That May Spread to the Great Lakes

Table B-1. Summary of Literature Review for Potentially Invasive Species. The 58 species Shown as Shaded Represent Those with the Reatest Risk

Type of Organism Common Name (Scientific Name)	In GL?[a]	Invasion History[b]	Spread Potential in GL[c]	Ecol. Impact in GL[d]	Species Origin[e]	Possible Source[f]	Concern, Consequence, Invasion[g]	Lit. Source[h]
Cnidarian Hydroid *(Cordylophora caspia)*	Yes	Yes	High	Med	Ponto-Caspian Sea	aquarium; ballast water	established in L. Erie, Baltic Sea basin; invasion history	13, 19, 52
crustacean Amphipod *(Pontogammarus crassus)*	No	Yes	NEK	NEK	Caspian Sea	ballast water	established in Baltic Sea; not established yet, may be introduced	13,19, 30
crustacean Amphipod *(Corophium curvispinum)*	No	Yes	High	Med	Caspian Sea	ballast water	found in L. St. Clair; invasion history; displaced filter-feeding caddisflies widely distributed in Europe; high densities in Baltic ports; established itself in British Isles	19, 23, 29, 30
crustacean Amphipod *(Dikerogammarus villosus)*	No	Yes	High	High	Ponto-Caspian Sea	ballast water; canals (Europe)	not established yet; established in Baltic Sea basin; eat and replace native amphipods; invading and decreasing natives in Europe; have rapid impact on macroinvertebrate survival, leading to population declines; invading most of Western Europe's hydrosystems	6, 7, 8, 13, 19, 30
crustacean Amphipod *(Pontogammarus robustoides)*	No	Yes	High	NEK	Ponto-Caspian Sea	ballast water; canals (Europe)	not established yet; established in Baltic Sea basin	13, 19, 30

Table B-1. (Continued)

Type of Organism Common Name (Scientific Name)	In GL?[a]	Invasion History[b]	Spread Potential in GL[c]	Ecol. Impact in GL[d]	Species Origin[e]	Possible Source[f]	Concern, Consequence, Invasion[g]	Lit. Source[h]
crustacean Amphipod *(Echinogammarus ischnus)*	Yes	Yes	High	Med	Ponto-Caspian Sea	ballast water	established in L. Erie, Huron, Ontario; invasion history; limited dispersal capability; displaced native amphipod *Gmelinoides fasciatus*	4, 9, 13, 19, 25, 26, 29, 32, 52, 53
crustacean Amphipod *(Dikerogammarus haemobaphes)*	No	Yes	High	Med	Ponto-Caspian Sea	ballast water; canals (Europe)	not established yet; established in Baltic Sea basin	13, 19, 30
crustacean Amphipod *(Corophium sowinskyi)*	No	Yes	Med	Med	Ponto-Caspian Sea	canals	spreading across Europe; not establish-hed yet; could alter littoral communi-ties and food webs	30
crustacean Amphipod *(Echinogammarus warpachowskyi)*	No	Yes	High	Med	Ponto-Caspian Sea	ballast water	not established yet; established in Baltic Sea basin	*13, 19*
crustacean Amphipod *(Pontogammarus aralensis)*	No	Yes	High	Med	Ponto-Caspian Sea	ballast water	not established yet; established in Baltic Sea basin	*13*
crustacean Amphipod *(Echinogammarus berilloni)*	No	NEK	NEK	NEK	Ponto-Caspian Sea	ballast water	not established yet, may be introduced	*13*
crustacean Amphipod *(Echinogammarus trichiatus)*	No	NEK	NEK	NEK	Ponto-Caspian Sea	ballast water	not established yet, may be introduced	*13*
crustacean Amphipod *(Gammarus tigrinus)*	No	NEK	NEK	NEK	North America	ballast water	not established yet, may be introduced	*13*

Table B-1. (Continued)

Type of Organism Common Name (Scientific Name)	In GL?[a]	Invasion History[b]	Spread Potential in GL[c]	Ecol. Impact in GL[d]	Species Origin[e]	Possible Source[f]	Concern, Consequence, Invasion[g]	Lit. Source[h]
crustacean Amphipod *(Gmelinoides fasciatus)*	No	NEK	NEK	NEK	Asia	ballast water	not established yet, may be introduced	*13*
crustacean Amphipod *(Iphigenella shablensis)*	No	NEK	NEK	NEK	Ponto-Caspian Sea	ballast water	not established yet, may be introduced	*13*
crustacean Amphipod *(Pontogammarus maeoticus)*	No	NEK	NEK	NEK	Ponto-Caspian Sea	ballast water	not established yet, may be introduced	*13*
crustacean Amphipod *(Pontogammarus obesus)*	No	NEK	NEK	NEK	Ponto-Caspian Sea	ballast water	not established yet, may be introduced	13, 30
crustacean Amphipod *(Pontogammarus subnudas)*	No	NEK	NEK	NEK	Ponto-Caspian Sea	ballast water	not established yet, may be introduced	13
crustacean Isopod *(Jaera istri)*	No	NEK	NEK	NEK	Ponto-Caspian Sea	ballast water	not established yet, may be introduced	13
crustacean Isopod *(Jaera sarsi)*	No	NEK	NEK	NEK	Ponto-Caspian Sea	ballast water	not established yet, may be introduced	13
crustacean Isopod *(Proasellus coxalis)*	No	NEK	NEK	NEK	Ponto-Caspian Sea	ballast water	not established yet, may be introduced	13
crustacean Isopod *(Proasellus meridianus)*	No	NEK	NEK	NEK	Ponto-Caspian Sea	ballast water	not established yet, may be introduced	13

Table B-1. (Continued)

Type of Organism Common Name (Scientific Name)	In GL?[a]	Invasion History[b]	Spread Potential in GL[c]	Ecol. Impact in GL[d]	Species Origin[e]	Possible Source[f]	Concern, Consequence, Invasion[g]	Lit. Source[h]
crustacean Mysid shrimp *(Hemimysis anomala)*	No	Yes	High	Med	Ponto-Caspian Sea	ballast water	not established yet; established in Baltic Sea basin; reduction of zooplankton; biomagnification of contaminants at higher trophic levels; adaption to shallow, warm waters	13, 19, 30
crustacean Mysid shrimp *(Paramysis ullskyi)*	No	Yes	Med	NEK	Ponto-Caspian Sea	ballast water	not established yet; established in Baltic Sea basin; reduction of zooplankton; biomagnification of contaminants at higher trophic levels; adaption to shallow, warm waters	30
crustacean Mysid shrimp *(Limnomysis benedeni)*	No	Yes	NEK	NEK	Ponto-Caspian Sea	ballast water	not established yet; established in Baltic Sea basin; reduction of zooplankton; biomagnification of contaminants at higher trophic levels; adaption to shallow, warm waters	13, 19, 30
crustacean Mysid shrimp *(Paramysis intermedia)*	No	Yes	NEK	NEK	Ponto-Caspian Sea	ballast water	not established yet; established in Baltic Sea basin; reduction of zooplankton; biomagnification of contaminants at higher trophic levels; adaption to shallow, warm waters	13, 30
crustacean Mysid shrimp *(Paramysis lacustris)*	No	Yes	NEK	NEK	Ponto-Caspian Sea	ballast water	not established yet; established in Baltic Sea basin; reduction of zooplankton; biomagnification of contaminants at higher	13, 19, 30

Table B-1. (Continued)

Type of Organism Common Name (Scientific Name)	In GL?[a]	Invasion History[b]	Spread Potential in GL[c]	Ecol. Impact in GL[d]	Species Origin[e]	Possible Source[f]	Concern, Consequence, Invasion[g]	Lit. Source[h]
							trophic levels; adaption to shallow, warm waters	
crustacean Pseudocu-mid *(Pseudocuma cercaroides)*	No	NEK	NEK	NEK	Ponto-Caspian Sea	ballast water	not established yet, may be introduced	13
crustacean Pseudocumid *(Pterocuma pectinata)*	No	NEK	NEK	NEK	Ponto-Caspian Sea	ballast water	not established yet, may be introduced	13
crustacean Rusty crayfish *(Orconectes rusticus)*	No	Yes	High	High	Southeas t United States	bait	found in inland waters of Michigan (not Great Lakes); invasion history; spreading north into Ontario, northern Midwest; reduction of vegetation important to native fish for food and cover; habitat destruction; affects native crayfish industry (Ontario)	4, 17, 46, 48, 49, 56
fish Alewife *(Alosa pseudoharengus)*	Yes	Yes	Present	High	Atlantic coast of North America	bait fish; canals	found in L. Ontario, Erie, Michigan, Superior; restructure a lake's food web, leaving less food for native fish; contributed to extinction of some native species	5, 9, 11, 29, 30, 35, 44, 49, 52, 55, 56
fish Arowana *(Osteoglossum bicirrhosum)*	No	NEK	NEK	NEK	South America	ornamental	potential there but have slow growth rate; inability to survive temperatures below 58 degrees Fahrenheit	11, 31

Table B-1. (Continued)

Type of Organism Common Name (Scientific Name)	In GL?[a]	Invasion History[b]	Spread Potential in GL[c]	Ecol. Impact in GL[d]	Species Origin[e]	Possible Source[f]	Concern, Consequence, Invasion[g]	Lit. Source[h]
fish Bighead carp *(Hypophthalmichthys nobilis / Aristichthys nobilis)*	Yes	Yes	High	Med	China	aquaculture; fish market; canals	few isolated cases in L. Erie; tolerate low temp; established in Mississippi River basin; invasion history; vast mobility; voracious consumption habits; clog fishing nets and scare away commercial fish; competes with native fish	11, 12, 14, 17, 31, 34, 39, 43, 44, 46, 49, 56
fish Black carp *(Mylopharyngodon piceus)*	No	No	NEK	NEK	China	aquaculture	not established yet; reports found in conclave of Ohio, Missouri, and Mississippi Rivers; vast mobility; voracious consumption habits; clog fishing nets and scare away commercial fish; risk to commercial shellfish stocks	11, 18, 34, 39, 43, 46
fish Black sea silverside *(Aphanius boyeri)*	No	NEK	High	NEK	Ponto-Caspian Sea	bait; aquaculture		18
fish Bleak *(Alburnus alburnus)*	No	Yes	High	Med	Ponto-Caspian Sea	aquaculture; bait	invasive in Baltic; feed on crustaceans	11,18
fish Blue catfish *(Ictalurus furcatus)*	No	Yes	Med	Med	North America	aquaculture; sport fish	spreading outside natural range; will eat any species of fish they can catch, along with crayfish, freshwater mussels, frogs, and other readily available aquatic food sources	11, 18

Table B-1. (Continued)

Type of Organism Common Name (Scientific Name)	In GL?[a]	Invasion History[b]	Spread Potential in GL[c]	Ecol. Impact in GL[d]	Species Origin[e]	Possible Source[f]	Concern, Consequence, Invasion[g]	Lit. Source[h]
fish Blue tilapia *(Oreochromis aureus)*	No	Yes	NEK	NEK	Africa/ Eurasia	aquaculture; sport fish	spreading in U.S.; invasive in FL; local abundance and high densities in certain areas have resulted in marked changes in fish community structure; private culture in Ontario	11,17, 18, 42, 52
fish Blueback herring *(Alosa aestivalis)*	Yes	No	Present	High	Atlantic coast of North America	canals	invasive in L. Ontario; impede recovery of depressed populations of indigenous fishes such as cisco and lake trout; cold water may prevent establishment	9, 11, 28, 29, 52, 53
fish Bullhead *(Cottus gobio)*	No	NEK	High	NEK	Europe	bait	often behave aggressively towards one another, and competition for shelter and foraging space can be intense	11, 18
fish Caspian shad *(Caspialosa caspia)*	No	NEK	Med	NEK	Ponto-Caspian Sea	fish market		11, 18
fish Caucasian goby *(Knipowitschia caucasica)*	No	Yes	High	Med	Ponto-Caspian Sea	ballast water	invasion in Baltic; predators which feed on small benthic animals	11, 18
fish Cherry salmon *(Oncorhynchus masou)*	No	NEK	Med	NEK	North Pacific; Japan	sport fish	widespread	11, 18
fish Chub *(Leucaspius cephalus)*	No	NEK	NEK	NEK	Ponto-Caspian Sea	ornamental; live bait; sport fish		11, 18

Table B-1. (Continued)

Type of Organism Common Name (Scientific Name)	In GL?[a]	Invasion History[b]	Spread Potential in GL[c]	Ecol. Impact in GL[d]	Species Origin[e]	Possible Source[f]	Concern, Consequence, Invasion[g]	Lit. Source[h]
fish Chum salmon *(Oncorhynchus keta)*	No	NEK	High	NEK	Asia/ North Pacific	sport fish	widespread	11, 18
fish Clown loach *(Botia macracanthus)*	No	NEK	NEK	NEK	Indonesia	ornamental	potential there but have slow growth rate	11, 31
fish Common dace *(Leuciscus leuciscus)*	No	Yes	High	NEK	Ponto-Caspian Sea	bait	widespread in Europe, showed invasion in one country	11, 18
fish Common tilapia *(Oreochromis mossambica)*	No	NEK	NEK	NEK	Africa	fish market; aquaculture	private culture in Ontario	11, 17, 18, 42
fish Eurasian minnow *(Phoxinus phoxinus)*	No	Yes	High	NEK	Ponto-Caspian Sea	bait	invasive in Baltic	11, 18
fish European perch *(Perca fluviatilis)*	No	Yes	High	Med	Ponto-Caspian Sea	sport fish	widespread prized for angling (introduced in many countries); because widespread, designated as invasive due to impacts on native species; cannibalism is common	11, 18, 49
fish European ruffe *(Gymnocephalus cernuus)*	Yes	Yes	Present	High	Ponto-Caspian Sea	ballast water; canals; bait	invaded L. Superior, Huron; now in L. Michigan; invasive history in Europe; displace native species; predation on native fish eggs; competition with native fish; L. Superior native fish declined since introduction of ruffe	11, 16, 23, 29, 30, 37, 43, 16, 48, 49, 52, 56

Table B-1. (Continued)

Type of Organism Common Name (Scientific Name)	In GL?[a]	Invasion History[b]	Spread Potential in GL[c]	Ecol. Impact in GL[d]	Species Origin[e]	Possible Source[f]	Concern, Consequence, Invasion[g]	Lit. Source[h]
fish European whitefish (Vendace) *(Core gonus albula)*	No	Yes	High	NEK	Europe	sport fish	some invasion history	11, 18
fish Fourspine stickleback *(Apeltes quadracus)*	Yes	NEK	NEK	NEK	Europe/ Atlantic coast of North America	ballast water; ornamental	confined to coasts; rapid increases of *Apeltes* in Thunder Bay (L. Superior) suggests the species is displacing native sticklebacks at a rapid rate	11, 52
fish Ghost (or glass) catfish *(Kryptopterus bicirrhis)*	No	NEK	NEK	NEK	Asia	ornamental	potential there but have slow growth rate	11, 31
fish Giant or red snakehead *(Channa micropeltes)*	No	NEK	NEK	NEK	Asia	ornamental	potential there but have slow growth rate	11, 31
fish Goldfish *(Carassius auratus)*	Yes	Yes	High	High	China/ Japan	ornamental	of the large established populations are recorded from the vicinity of western L. Erie, widespread; extensive invasion history; concern of impacts on community, increasing turbidity, predation of native fish, help facilitate algal blooms; tolerate low temperatures	9, 11, 31, 49
fish Grass carp *(Ctenopharyngodon idella)*	No	Yes	High	High	Eastern Asia	aquaculture; fish market	found isolated in L. Erie, Ontario, Huron; grazes on aquatic vegetation reducing plant density or removing all	11, 14, 17, 31, 43, 44, 46, 52

Table B-1. (Continued)

Type of Organism Common Name (Scientific Name)	In GL?[a]	Invasion History[b]	Spread Potential in GL[c]	Ecol. Impact in GL[d]	Species Origin[e]	Possible Source[f]	Concern, Consequence, Invasion[g]	Lit. Source[h]
							aquatic vegetation from a body of water; competes with native fish; tolerate low temperatures; invasion history	
fish Inland silverside *(Menidia beryllina)*	No	Yes	NEK	NEK	Eastern U.S. coast	aquaculture; sport fish	could replace native fish; replaced native fish in CA, OK; found in Mississippi conclave of Illinois and Ohio Rivers	11, 18, 52
fish Koi (common) carp *(Cyprinus carpio)*	Yes	Yes	High	High	Eurasia	ornamental; aquaculture	found in Great Lakes; extensive invasion history; globally widespread; uproot and destroy submerged aquatic vegetation and therefore may be detrimental to duck and native fish populations; tolerate low temperatures	11, 31, 49
fish Longtail goby *(Ctenogobius sagittula)*	No	NEK	NEK	NEK	Eastern Pacific			11, 18
fish Monkey goby *(Neogobius fluviatilis)*	No	Yes	High	High	Eurasia	ballast water	invasion history in Europe; mass invasion of monkey goby is connected with the intensive consumption of plankton crustaceans; competes with other small fish for food and space	11, 14, 18, 30, 51

Table B-1. (Continued)

Type of Organism Common Name (Scientific Name)	In GL?[a]	Invasion History[b]	Spread Potential in GL[c]	Ecol. Impact in GL[d]	Species Origin[e]	Possible Source[f]	Concern, Consequence, Invasion[g]	Lit. Source[h]
fish Mummichog *(Fundulus heteroclitus)*	No	NEK	NEK	NEK	Western Atlantic/ Eastern U.S.coast	bait; aquarium		11, 18
fish Nile tilapia *(Oreochromis niloticus)*	No	NEK	NEK	NEK	Africa	fish market; aquaculture	private culture in Ontario; Mississippi	11, 7,18, 42
fish Oriental weatherfish *(Misgurnus anguillicaudatus*	Yes	Yes	High	Med	Asia	ornamental; aquaculture	established in Shiawassee River and L. Michigan; reduce populations of aquatic insects important as food to native fishes; invasion history; tolerate low temperatures	11, 14, 31, 44
fish Pacu *(Coloss-oma macropomum)*	No	NEK	NEK	NEK	South America	aquaculture; ornamental	may compete with larvae of native fish species for plankton	11,18
fish Pike killifish *(Belonesox belizanus)*	No	NEK	NEK	NEK	Central America	aquarium	voracious predator and has been known to reduce populations of eastern mosquitofish (*Gambusia holbrooki*) and other native poeciliid and cyprinodontid populations	11,18
fish Pontic shad *(Alosa pontica)*	No	NEK	Med	NEK	Ponto-Caspian Sea	fish market; aquarium		11,18
fish Racer goby *(Neogobius gymnotrachelus*	No	Yes	High	Med	Ponto-Caspian Sea	canals; ballast water	spreading in Europe	11,18, 51

Table B-1. (Continued)

Type of Organism Common Name (Scientific Name)	In GL?[a]	Invasion History[b]	Spread Potential in GL[c]	Ecol. Impact in GL[d]	Species Origin[e]	Possible Source[f]	Concern, Consequence, Invasion[g]	Lit. Source[h]
fish Rainbow smelt *(Osmerus mordax)*	Yes	Yes	Med	High	Atlantic coast of Central America	sport fish	Erie Canal; Lake Ontario, Michigan, Superior; Minnesota; dominant prey form for salmonids; contributed to extinction of blue pike; affect imperiled species	11, 44, 52, 53, 55, 56
fish Red tail botia *(Botia modesta)*	No	NEK	NEK	NEK	Asia	ornamental	potential there but have slow growth rate	11, 31
fish Redear sunfish *(Lepomis microlophus)*	Yes	NEK	High	NEK	Atlantic coast of North America	aquaculture; sport fish	found in L. Michigan; introduced redear are associated with ecological changes in populations of pumpkinseed *L. gibbosus*, a native molluscivore; preference for mollusks	11, 18, 52
fish Roach *(Rutilus rutilus)*	No	Yes	High	Med	Ponto-Caspian Sea	sport fish	invasive in Baltic; nuisance once established	11, 18
fish Round goby *(Neogobius melanostromus)*	Yes	Yes	Present	High	Ponto-Caspian Sea	ballast water	found in Great Lakes; invasion history; spreads rapidly; zebra mussels domin-ate diet; altering benthic communities; aggressive; lake trout egg, sturgeon egg predators	3, 9, 11, 29, 30, 32, 43, 46, 48, 49, 51, 52, 53, 56
fish Rudd *(Scardin-iuserythrophthalmus)*	Yes	Yes	Present	High	Eurasia	bait fish; ballast water	introduced into L. Ontario; expanded in L. Erie; habitat degradation for native fish; rudd introduced to open waters will hybridize with golden shiner	11, 17, 29, 41, 46, 49

Table B-1. (Continued)

Type of Organism Common Name (Scientific Name)	In GL?[a]	Invasion History[b]	Spread Potential in GL[c]	Ecol. Impact in GL[d]	Species Origin[e]	Possible Source[f]	Concern, Consequence, Invasion[g]	Lit. Source[h]
fish Sand goby *(pomatoschistus minutus)*	No	Yes	High	Med	Eastern Atlantic/ Ponto-Caspian Sea	aquarium; aquaculture	abundant; invasion in Baltic; predators that feed on small benthic animals	11, 18
fish Silver carp *(Hypophthalmichthys molitrix)*	No	NEK	High	NEK	China	aquaculture	documented in Cal-Sag Channel of IL Waterway; spread along Mississippi River; competition with commercially significant native fish for food and habitat; vast mobility; voracious consumption habits; clog fishing nets; widespread in U.S. Gulf states	11, 12, 18, 34, 39, 43, 44, 46, 52, 56
fish Snakehead *(Channa argus argus)*	No	Yes	High	Med	Asia	fish market	sporadic places; devastation to freshwater ecosystems of the U.S. because of its predacious nature, lack of natural predators, high fertility, and adaptability to a wide range of environmental conditions	11, 12, 46, 49
fish Starry goby *(Benthophilus stellatus)*	No	Yes	High	NEK	Ponto-Caspian Sea	ballast water?	spreading in Europe	11, 18, 30
fish Striped bass *(Morone saxatilis)*	No	Yes	NEK	NEK	Atlantic coast of North America	fish market	invasion history	11, 31

Table B-1. (Continued)

Type of Organism Common Name (Scientific Name)	In GL?[a]	Invasion History[b]	Spread Potential in GL[c]	Ecol. Impact in GL[d]	Species Origin[e]	Possible Source[f]	Concern, Consequence, Invasion[g]	Lit. Source[h]
fish Sunbleak *(Leucaspius delineatus)*	No	Yes	High	High	Ponto-Caspian Sea	canals	widespread in Europe; affecting coarse fish populations in England (limited range); feeds on phytoplankton and zooplankton	11, 18
fish Tench *(Tinca tinca)*	No	Yes	High	High	Ponto-Caspian Sea/ Eur-ope/Asia	ornamental; canals; sport fish	been captured in St. Lawrence River; may head to L. Ontario; competition with benthic feeders, native fish; potential competitor for food with sport fishes and native cyprinids; widespread	11, 17, 18, 37, 49
fish Toothed carp *(Aphanius fasciatus)*	No	NEK	High	NEK	Ponto-Caspian Sea	canal	competes with native fish; declining in Mediterranean	11, 18
fish Tubenose goby *(Proterorhinus marmoratus)*	Yes	Yes	Med	High	Ponto-Caspian Sea	ballast water	found/restricted to L. St. Clair, western L. Erie; Invasion history but not spread rapidly; aggressive	11, 29, 32, 43, 51, 52
fish Tyulka / Caspian kilka *(Clupeonella cultriventris/caspia)*	No	Yes	High	NEK	Ponto-Caspian Sea	ballast water	expanded in Volga River; suppressed native fish	11, 14, 18, 30
fish Ukranian or nine-spined stick-leback *(Pungitius platygaster)*	Yes	NEK	Present	NEK	Ponto-Caspian Sea	canal?; bait?	found in L. Huron, Michigan	11, 18, 52

Table B-1. (Continued)

Type of Organism Common Name (Scientific Name)	In GL?[a]	Invasion History[b]	Spread Potential in GL[c]	Ecol. Impact in GL[d]	Species Origin[e]	Possible Source[f]	Concern, Consequence, Invasion[g]	Lit. Source[h]
fish Weather fish *(Misgurnus fossiis)*	No	Yes	NEK	NEK	Europe	ornamental	tolerate low temperature; established invasion in Italy, Spain, Croatia	11, 31
fish White cloud mo-untain minnow *(Ta-nichthys albonubes)*	No	Yes	High	Med	China	ornamental	tolerate low temperature; extensive invasion history; high occurrence frequency	11, 14, 31
fish White perch *(Morone americana)*	Yes	Yes	Present	High	Atlantic coast of North America	fish market; canals	already invaded Great Lakes; competition with native fish; potential to cause declines of walleye populations; prey on eggs of walleye	9, 11, 17, 31, 44, 46, 48, 52, 53, 56
fish Zander *(Sander lucioperca)*	No	Yes	High	High	Ponto-Caspian Sea	sport fish; canals; aquaculture	found everywhere in Europe; depleted stocks of native fish; hunts in packs	11, 18
flatworm Flatworm *(Dendrocoelum romanodanubiale)*	No	NEK	NEK	NEK	North Sea basin	ballast water	not established yet, may be introduced	13
flatwormTrematode *(Apophallus muehlingi)*	No	NEK	NEK	NEK	Black Sea	ballast water	not established yet, may be introduced	13
flatworm Trematode *(Nicolla skrjabini)*	No	NEK	NEK	NEK	Black Sea	ballast water	not established yet, may be introduced	13
flatworm Trematode *(Rossicotrema donicum)*	No	NEK	NEK	NEK	Black Sea	ballast water; fish host	not established yet, may be introduced	13
insect Apterogote (wingless) *Campodea staphylinus)*	No	Yes	High	NEK	Eurasia	ballast water	widespread; been recorded in North American fresh waters; tolerate range of salinity	10

Table B-1. (Continued)

Type of Organism Common Name (Scientific Name)	In GL?[a]	Invasion History[b]	Spread Potential in GL[c]	Ecol. Impact in GL[d]	Species Origin[e]	Possible Source[f]	Concern, Consequence, Invasion[g]	Lit. Source[h]
microcrustacean Baltic water flea *(Bosmina coregoni)*	Yes	Yes	Present	High	Eurasia/ Ponto-Caspian	ballast water	established in L. Ontario, Superior; invasion history	13
microcrustacean Calanoid copepod *(Eurytemora affinis)*	Yes	Yes	Present	NEK	Ponto-Caspian Sea	ballast water	widespread; invasion history; established in Great Lakes; high populations	13
microcrustacean Calanoid copepod *(Heterocope appendiculata)*	No	Yes	High	NEK	Ponto-Caspian Sea	ballast water	established in Baltic Sea basin	13
microcrustacean Calanoid copepod *(Calanipeda aquae-dulcis)*	No	Yes	High	NEK	Ponto-Caspian Sea	ballast water	established in Baltic Sea basin	13
microcrustacean Calanoid copepod *(Heterocope caspia)*	No	Yes	High	NEK	Ponto-Caspian Sea	ballast water	established in Baltic Sea basin	13
microcrustacean Cladoceran *(Daphnia cristata)*	No	Yes	High	NEK	Ponto-Caspian Sea	ballast water	established in Baltic Sea basin	13
microcrustacean Cladoceran *(Bosmina obtusirostris)*	No	Yes	High	NEK	Ponto-Caspian Sea	ballast water	established in Baltic Sea basin	13
microcrustacean Cladoceran *(Podonevadne trigona ovum)*	No	Yes	High	NEK	Ponto-Caspian Sea	ballast water	established in Baltic Sea basin	13

Table B-1. (Continued)

Type of Organism Common Name (Scientific Name)	In GL?[a]	Invasion History[b]	Spread Potential in GL[c]	Ecol. Impact in GL[d]	Species Origin[e]	Possible Source[f]	Concern, Consequence, Invasion[g]	Lit. Source[h]
microcrustacean Cladoceran (water flea) *(Daphnia lumholtzi)*	No	Yes	Med	NEK	Australia; SE Asia	ballast water; boating; waterfowl	expanding in U.S.; widespread; prefers warm water; competes with native daphnia for food and of its ability to avoid predation	14, 34, 52, 53
microcrustacean Cladoceran *(Cor-nigerius maeoticus maeoticus)*	No	Yes	High	NEK	Ponto-Caspian Sea	ballast water	established in Baltic Sea basin	13
microcrustacean Cyclopoid copepod *(Cyclops strenuus)*	Yes	Yes	Present	NEK	Eurasia	ballast water	established in Great Lakes	10
microcrustacean Cyclopoid copepod *(Cyclops kolensis)*	No	Yes	High	NEK	Ponto-Caspian Sea	ballast water	established in Baltic Sea basin	13
microcrustacean Fishhook waterflea *(Cercopagis pengoi)*	Yes	Yes	Present	High	Ponto-Caspian Sea	ballast water; boating	established in Great Lakes except L. Huron, Superior; Baltic Sea basin; compete with small fish for zooplan-kton; invasive in Europe; clogs nets for fisheries; damage cost data available	9, 13, 19, 20, 29, 32, 38, 49, 52, 53, 54, 56
microcrustacean Harpactacoid copepod *(Ectinosoma abrau)*	No	NEK	High	NEK	Ponto-Caspian Sea	ballast water	potential to invade Great Lakes	13
microcrustacean Harpactacoid cope-pod *(Onychocamptus mohammed)*	Yes	NEK	Present	NEK	Ponto-Caspian Sea	ballast water	established in L. Ontario	13, 14

Table B-1. (Continued)

Type of Organism Common Name (Scientific Name)	In GL?[a]	Invasion History[b]	Spread Potential in GL[c]	Ecol. Impact in GL[d]	Species Origin[e]	Possible Source[f]	Concern, Consequence, Invasion[g]	Lit. Source[h]
microcrustacean Harpactacoid cope-pod *(Paraleptastacus spinicaudata triseta)*	No	Yes	High	NEK	Ponto-Caspian Sea	ballast water	invasion history	13
microcrustacean Harpactacoid cope-pod *(Schizopera borutzkyi)*	Yes	Yes	Present	High	Danube River delta of Black Sea	ballast water	established in L. Michigan; altering species composition of nearshore communities	1, 10, 13, 14, 15, 28
microcrustacean Harpactacoid cope-pod *(Nitocra incerta)*	Yes	Yes	Present	NEK	Black Sea	ballast water	established in L. Michigan	13
microcrustacean Spiny waterflea *(Bythotrephes longimanus)*	Yes	Yes	High	High	Great Britain; Europe	ballast water; boating	established in Great Lakes; Great Lakes - has caused major changes in the zoop-lankton community structure; invasion history; reproduce rapidly; competes directly with small fish and can have impact on zooplankton community; damage cost data available	13, 21, 33, 38, 48, 49, 52, 53, 54
microcrustacean Spiny waterflea *(Bythotrephes cederstroemi)*	Yes	Yes	Present	High	Eurasia	ballast water	established in Great Lakes; the invasion into the Laurentian Great Lakes has resulted in substantial and sustained decreases in the populations of a number of (mostly cladoceran) native zooplankton species; coincided with	32, 49, 50, 56

Table B-1. (Continued)

Type of Organism Common Name (Scientific Name)	In GL?[a]	Invasion History[b]	Spread Potential in GL[c]	Ecol. Impact in GL[d]	Species Origin[e]	Possible Source[f]	Concern, Consequence, Invasion[g]	Lit. Source[h]
							dramatic declines in abundance of Daphnia	
microsporidian Fish parasite *Heterosp-oris sp.)*	Yes	Yes	NEK	NEK	Europe/ Asia ?	fish release; bait	found in L. Ontario, attacks muscle ce-lls in yellow perch; found in Wisconsin	9, 46
mollusk Asian clam *(Corbicula fluminea)*	Yes	Yes	NEK	NEK	Asia	ornamental; fish market/bait	established in L. Michigan, Superior, Erie; does not tolerate low temperatu-res; fouled water plants; damage cost data available	13, 46, 49, 52, 54
mollusc Basket (European) shell *(Corbula gibba)*	No	NEK	NEK	NEK	Atlantic coast of Europe	ballast water	capacity to achieve very high population densities, giving it the potential to affect the growth and recruitment of a wide range of soft bottom organisms	13
mollusk Clam *(Hypanis colorata)*	No	NEK	NEK	NEK	Caspian Sea	ballast water	potential to invade Great Lakes	13, 30
mollusc European fingernail clam *(Sphaerium corneum)*	Yes	Yes	Present	NEK	Eurasia	unknown - canals?	found in L. Huron, Ontario; effect unknown	52
mollusc Golden mussel *(Limnoperna fortunei)*	No	Yes	Med	Med	China/ Asia	ballast water	produces a rapid change in benthic communities and threatens native bio-diversity; produces macrofouling in the water systems of facilities; spreading in South America	27, 49

Table B-1. (Continued)

Type of Organism Common Name (Scientific Name)	In GL?[a]	Invasion History[b]	Spread Potential in GL[c]	Ecol. Impact in GL[d]	Species Origin[e]	Possible Source[f]	Concern, Consequence, Invasion[g]	Lit. Source[h]
mollusc Chinese mystery snail [(prosobranch] *(Cipangopaludina chinensis malleata)*	Yes	Yes	High	Med	Asia	fish market; ornamental	isolated pop. in L. Erie and upper St. Lawrence River; established L. Michi-gan; clog screens of water intake pipes; vectors for the transmission of parasites and diseases	31, 42, 46, 52
mollusc New Zealand mudsnail [hyd-robid] *(Potamop-yrgus antipodarum)*	Yes	Yes	Present	High	New Zealand	ballast water	established in L. Ontario, Superior; invasion history; reduce native species, harm trout populations; suspected that can alter primary production of streams; spread rapidly	13, 23, 28, 44, 46, 49, 52
mollusc Quagga mussel *(Dreissena bugensis)*	Yes	Yes	Present	Med	Ukraine/ Ponto-Caspian Sea	ballast water	found in L. Erie, Ontario; invasion history; historically dominated the biomass of transition(30-50m) and profundal regions(>50m); negative impact on raw water-using industries, potable water treatment plants; damage cost data available	9, 26, 29, 30, 44, 49, 52, 54, 56
mollusc Snail (hydro-bid) *(Lithoglyphus naticoides)*	No	NEK	NEK	NEK	Ponto-Caspian	ballast water	potential to invade Great Lakes	13
mollusc Snail (neritid) *(Theodoxus fluviatiis)*	No	NEK	NEK	NEK	Baltic/ Black Sea	ballast water	potential to invade Great Lakes	13
mollusc Snail (neritid) *(Theodoxus pallasi)*	No	NEK	NEK	NEK	Caspian Sea	ballast water	potential to invade Great Lakes	13

Table B-1. (Continued)

Type of Organism Common Name (Scientific Name)	In GL?[a]	Invasion History[b]	Spread Potential in GL[c]	Ecol. Impact in GL[d]	Species Origin[e]	Possible Source[f]	Concern, Consequence, Invasion[g]	Lit. Source[h]
mollusc Zebra mussel *(Dreissena polymorpha)*	Yes	Yes	Present	High	Ponto-Caspian Sea	ballast water	already established in Great Lakes; invasion history; interferes with native molluscs' ecological functions; damage cost data available	2, 9, 29, 32, 44, 48, 49, 52, 53, 56
plant Curly-leaf pondweed *(Potamo-geton crisp us)*	Yes	Yes	Present	Med	Eurasia	ornamental; horticulture	established in U.S.; invasion history; causes problems due to excessive growth	22, 41, 47, 48, 49, 56
plant Dwarf hygrophila *(Hygrophila polysperma)*	No	Yes	High	Med	India	ornamental	tolerate low temp; established in FL, TX, VA; invasion history; clogs irrigation and flood-control canals	14, 31
plant Brazilian elodea or waterweed *(Egeria densa)*	No	Yes	Med	High	Central/ South America	ornamental	established in OR, NY, MD, CT; invasion history; slow dispersal; create dense mats that can impede water recreation; water movement is restricted causing fish population imbalances; cause fluctuations in water quality	14, 31, 36, 46, 49
plant Eurasian watermilfoil *(Myriophyllum spicatum)*	Yes	Yes	High	High	Eurasia	ornamental; boaters	invaded L Ontario; interferes with water recreation; canopy can crowd out important native plants; decrease oxygen levels when plant decays; rapid colonization; damage cost data available; threat factor to fish	9, 17, 23, 47, 48, 49, 54, 55, 56

Table B-1. (Continued)

Type of Organism Common Name (Scientific Name)	In GL?[a]	Invasion History[b]	Spread Potential in GL[c]	Ecol. Impact in GL[d]	Species Origin[e]	Possible Source[f]	Concern, Consequence, Invasion[g]	Lit. Source[h]
plant European frogbit *(Hydrocharis morus-ranae)*	Yes	Yes	Med	High	Eurasia	ornamental; horticulture; boaters	invaded L. Ontario (only lake found); creates dense mats of vegetation and thus prevents light and nutrients from reaching submerged vegetation; plants die in the fall so depleted dissolved oxygen is possible	9, 17, 22, 36, 41, 46, 49, 56
plant European water clover *(Marsilea quadrifolia)*	Yes	Yes	Med	Med	Eurasia	ornamental	found in L. Ontario; poses a realistic nuisance threat to ecosystems; affect local molluscan communities	36
plant Fanwort *(Cabomba caroliniana)*	Yes	Yes	Present	High	Southeast United States	ornamental	Already invaded L. Michigan; clogs drainage canals and freshwater streams; form dense stands crowding out previously well-established plants	17, 31, 36, 49,
plant Flowering rush *(Butomus umbellatus)*	Yes	Yes	Med	NEK	Eurasia	ornamental; horticulture	spread in limited areas in Great Lakes (Erie, Ontario, Michigan); actively expanding; competes with native shoreline vegetation	22, 46, 48, 49, 56
plant Giant salvinia *(Salvinia molesta)*	No	Yes	Med	Med	South America	ornamental; horticulture	established in lower U.S.; invasion history; impede the flow of water to irrigation pipes and other water intake pipes; rapidly expanding populations can overgrow and replace native	22, 41, 46, 49

Table B-1. (Continued)

Type of Organism Common Name (Scientific Name)	In GL?[a]	Invasion History[b]	Spread Potential in GL[c]	Ecol. Impact in GL[d]	Species Origin[e]	Possible Source[f]	Concern, Consequence, Invasion[g]	Lit. Source[h]
							plants,resulting dense surface cover prevents light and atmospheric oxygen from entering the water; decomposing material drops to the bottom, greatly consuming dissolved oxygen needed by fish and other aquatic life	
plant Hydrilla *(Hydrilla verticillata)*	No	Yes	High	High	Central Africa	ornamental; horticulture	established in southern U.S.; high movement; invasion history; adaptable; dense mats of hydrilla will alter the waters chemistry by raising pH, cause wide oxygen fluctuations, and increase water temperature; eliminate native plants	22, 36, 41, 46, 49, 56
plant Minor (slender) naiad *(Najas minor)*	Yes	Yes	NEK	NEK	Europe	ornamental	limited in L. Erie; can form dense, monospecific stands in shallow water and hinder swimming, fishing, boating, and other forms of water contact recreation	36,
plant Parrot's feather *(Myriophyllum aquaticum)*	No	Yes	High	High	South America	ornamental	invasion history; very adaptive to variety of environments	14, 31, 49
plant Purple loosestrife *(Lythrum salicaria)*	Yes	Yes	Present	High	Eurasia	ornamental; horticulture	established in Great Lakes; invasion history; plant can form dense, impenet-rable stands	22, 23, 41, 47, 53, 54, 56

Table B-1. (Continued)

Type of Organism Common Name (Scientific Name)	In GL?[a]	Invasion History[b]	Spread Potential in GL[c]	Ecol. Impact in GL[d]	Species Origin[e]	Possible Source[f]	Concern, Consequence, Invasion[g]	Lit. Source[h]
							which are unsuitable as cover, food, or nesting sites for a wide range of native wetland animals; damage cost data available	
Plant Sessile joyweed *(Alternanthera sessilis)*	No	Yes	Med	NEK	Asia	ornamental; horticulture	invasion history; aggressive	22
plant Spiny naiad *(Najas marina)*	Yes	Yes	Med	NEK	Eurasia	ornamental	found in L. Ontario; widespread; interferes with recreational boating; pose a realistic nuisance threat to ecosystems	36
plant Variable-leaved watermilfoil *(Myriophyllum heterophyllum)*	No	Yes	High	Med	Eastern North America	ornamental; boaters	spreading in New England states; out compete native aquatic vegetation, resulting in nearly monotypic growth with less habitat value; interferes with recreational boating; pose a realistic nuisance threat to ecosystems	31, 36, 40
plant Water primrose *(Ludwigia uruguayensis)*	No	NEK	NEK	NEK	South America	ornamental	large colonies can prevent small boat navigation and recreational use of shoreline areas	36
plant Waterchestnut *(Trapa natans)*	Yes	Yes	High	Med	Eurasia	ornamental; horticulture	released in L. Ontario; reproduce rapidly; habitat degradation through floating mats, hindering navigation of waters and inhibiting the growth	9, 36, 44, 46, 54, 56

Table B-1. (Continued)

Type of Organism Common Name (Scientific Name)	In GL?[a]	Invasion History[b]	Spread Potential in GL[c]	Ecol. Impact in GL[d]	Species Origin[e]	Possible Source[f]	Concern, Consequence, Invasion[g]	Lit. Source[h]
							of native aquatic plant species; damage cost data available	
plant Yellow floating heart *(Nymphoides peltata)*	No	Yes	High	Med	Eurasia	ornamental; horticulture	spreading in New England; grows in dense patches, excluding native species and even creating stagnant areas with low oxygen levels underneath the floating mats	36, 46, 49
protozoans Amoebae (testate) *(Psammono-biotus communis)*	Yes	Yes	NEK	NEK	Baltic Sea	ballast water	found in L. Ontario; newly discovered in Great Lakes	9, 24
protozoans Amoebae (testate) *(Psammono-biotus dziwnowii)*	Yes	Yes	NEK	NEK	Baltic Sea	ballast water	found in L. Ontario; newly discovered in Great Lakes	9, 24
protozoans Amoebae (testate) *(Psammono-biotus linearis)*	Yes	Yes	NEK	NEK	Baltic Sea (?)	ballast water	found in L. Ontario and other Great Lakes	9
virus Largemouth bass virus *(Iridoviridae family)*	Yes	Yes	High	Med	?	bait fish	extended range in southeast basin; spreading in L. Michigan; kills fish ommercial, native, invasive)	34, 45, 46
worm Oligochaete *(Potamothrix bedoti)*	Yes	Yes	NEK	NEK	Ponto-Caspian Sea	ballast water	established in Great Lakes; invasion history in Baltic	13
worm Oligochaete (tubificid) *(Potamothrix heuscheri)*	No	Yes	NEK	NEK	Ponto-Caspian Sea	ballast water	invasion history in Baltic; continuously dispersing to the west over Central Europe and to the northwest towards the Baltic Sea	13

Table B-1. (Continued)

Type of Organism Common Name (Scientific Name)	In GL?[a]	Invasion History[b]	Spread Potential in GL[c]	Ecol. Impact in GL[d]	Species Origin[e]	Possible Source[f]	Concern, Consequence, Invasion[g]	Lit. Source[h]
worm Oligochaete *(Potamothrix moldaviensis)*	Yes	Yes	NEK	NEK	Ponto-Caspian Sea	ballast water	established in Great Lakes; invasion history in Baltic	13
worm Oligochaete *(Potamothrix vejdovskyi)*	Yes	Yes	NEK	NEK	Ponto-Caspian Sea	ballast water	established in Great Lakes; invasion history in Baltic	13
worm Polychaete worm *(Hypania invalida)*	No	Yes	NEK	NEK	Ponto-Caspian Sea	ballast water	invasion history in Baltic; slow spread in Europe	13, 30
worm Polychaete worm *(Hypaniola kowalevskyi)*	No	Yes	NEK	NEK	Ponto-Caspian Sea	ballast water	invasion history in Baltic; slow spread in Europe	13

[a] Present in Great Lakes is scored "yes" if the organism has been reported for any of the Lakes.

[b] Invasion history is scored "yes" if the species is reported to have an invasion history/exotic in any other country of the papers cited for that species.

[c] Spread potential within the Great Lakes after introduction scored "High" or "Med" based on comments in two or more of the papers cited for that species, or NEK (not enough known).

[d] Ecological impact if a species becomes established in Great Lakes scored as "High" or "Med" if so reported in at least two of the papers cited for that species; else, it is scored as NEK (not enough known).

[e] Species origin notes the probable native area or region of occurrence for the species in question, as cited in the papers reviewed for the species.

[f] The known, suspected, or probable mechanism of introduction of the species prior invasion history is listed based on the papers cited and prior invasion history.

[g] Comments related to investigators' concerns about the consequences of invasion for the species in question are abstracted from papers cited for the species in question.

[h] Articles referenced here were identified on the basis of species names plus reference to "Great Lakes". Thus, the listed references do not include many papers that consider the biology of the organism in its native range.

Table B-2. References That Identify 156 Invasive Species as Described in Table B-1

Reference Number	References
1	Horvath, TG; Oberdick, EJ; Last, LL; et al. (2001) Invading species dominate the Harpacticoid fauna in the nearshore sands of southern Lake Michigan. NABS Annual Meeting, La Crosse WI (abstract)
2	Buchan, LAJ; Padilla, DK. (1999) Estimating the probability of long-distance overland dispersal of invading aquatic species. Ecol Appl 9(1):254-265.
3	Corkum, LD; Sapota, RM; Skora, KE. (2004) The Round Goby, *Neogobius malanostomus*, a fish invader on both sides of the Atlantic Ocean. Biol Invasions 6(2):173-181.
4	Cristescu, MEA; Witt, JDS; Grigorovich, IA; et al. (2004) Dispersal of the Ponto-Caspian amphipod *Echinogammarus ischnus*: invasion waves from the Pleistocene to the present. Heredity 92:197-203.
5	Daniels, RA. (2001) Untested assumptions: the role of canals in the dispersal of Sea Lamprey, Alewife, and other fishes in the eastern United States. Environ Biol Fishes 60(4):309-329.
6	Devin, S; Piscart, C; Beisel, JN; et al. (2003) Ecological traits of the amphipod invader *Dikerogammarus villosus* on a mesohabitat scale. Arch Hydrobiol 158(1):43–56.
7	Dick, JTA; Platvoet, D. (2000) Invading predatory crustacean *Dikerogammarus villosus* eliminates both native and exotic species. Proc Biol Sci 267(1447):977-983.
8	Dick, JTA; Platvoet,D; Kelly, DW. (2002) Predatory impact of the freshwater invader *Dikerogammarus villosus* (Crustacea: Amphipoda). Can J Fish Aquat Sci 59(6):1078-1084.
9	Duggan, IC; Bailey, SA; Colautti, RI; et al. (2003) Biological invasions in Lake Ontario: past, present, and future. In: Munawar, M, ed. State of Lake Ontario (SOLO) – Past, Present, and Future.
10	Duggan, IC; van Overdijk, DA; Bailey, SA; et al. (2005) Invertebrates associated with residual ballast water and sediments of cargo- carrying ships entering the Great Lakes. Can J Fish Aquat Sci 62:2463-2474.
11	Froese, R; Pauly, D, Eds. (2005) FishBase.World Wide Web electronic publication. www.fishbase.org, version (10/2005). [1 2/07/05]
12	Goehle, M. (2004) Keeping track of aquatic invasive species in the Great Lakes. ANS Update. 9(4): 1-2. Available online at http://www.glc.org/ans/ansupdate/pdf/2004/Update-02-04.pdf
13	Grigorovich, IA; Colautti, RI; Mills, EL; et al. (2003) Ballast-mediated animal introductions in the Laurentian Great Lakes: retrospective and prospective analyses. Can J Fish Aquat Sci 60:740-756.
14	Holeck, KT; Mills, EL; MacIsaac, HJ; et al. (2004). Bridging troubled waters: Biological invasions, transoceanic shipping, and the Laurentian Great Lakes. Bioscience 54(1 0):91 9-929.
15	Horvath, TG; Whitman, RL; Last, LL. (2001) Establishment of two invasive crustaceans (Copepoda: Harpacticoida) in the nearshore sands of Lake Michigan. Can J Fish Aquat Sci 58:1261-1264.

Table B-2. (Continued)

Reference Number	References
16	Keppner, SM; Busiahn, TR; McClain, J; et al. (1997) Surveillance for Ruffe in the Great Lakes – an overview. G Lakes Res Rev 3(1):17-26.
17	Kerr, SJ; Brousseau, CS; Muschett, M. (2005) Invasive aquatic species in Ontario: a review and analysis of potential pathways for introduction. Fisheries 30(7):21 -30.
18	Kolar, CS; Lodge, DM. (2002) Ecological predictions and risk assessment for alien fishes in North America. Science 298:1233-1236.
19	MacIsaac, H; Grigorovich, IA; Ricciardi, A. (2001) Reassessment of species invasions concepts: the Great Lakes basin as a model. Biol Invasions 3:405-416.
20	MacIsaac, HJ; Robbins, TC; Lewis, MA. (2002) Modeling ships' ballast water as invasion threats to the Great Lakes. Can J Fish Aquat Sci 59:1245-1256.
21	MacIsaac, HJ; Borbely, JVM; Muirhead, JR; et al. (2004) Backcasting and forecasting biological invasions of inland lakes. Ecol Appl 1 4(3):773-783.
22	Maki, K; Galatowitsch, S. (2004) Movement of invasive aquatic plants into Minnesota (USA) through horticultural trade. Biol Conserv 118:389-396.
23	Mills, EL; Leach, JH; Carlton, JT; et al. (1994) Exotic species and the integrity of the Great Lakes. BioScience 44(10):666-676.
24	Nicholls, KH. (2005) *Psammonobiotus dziwnowi* and *Corythionella georgiana*, two new freshwater sand-dwelling testate amoebae (Rhizopoda: filosea). Acta Protozool 44:271-278.
25	Palmer, ME; Ricciardi, A. (2004) Physical factors affecting the relative abundance of native and invasive amphipods in the St. Lawrence River. Can J Zool 82:1886-1 893.
26	Palmer, ME; Ricciardi, A. (2005) Community interactions affecting the relative abundances of native and invasive amphipods in the St. Lawrence River. Can J Fish Aquat Sci 62:1111-1118.
27	Ricciardi, A. (1998) Global range expansion of the Asian mussel *Limnoperna fortunei* (Mytilidae): Another fouling threat to freshwater systems. Biofouling 13(2):97-106.
28	Ricciardi, A. (2001) Facilitative interactions among aquatic invaders: is an "invasive meltdown" occurring in the Great Lakes? Can J Fish Aquat Sci 58:2513-2525.
29	Ricciardi, A; MacIsaac, HJ. (2000) Recent mass invasion of the North American Great Lakes by Ponto-Caspian species. TREE 1 5(2):62-65.
30	Ricciardi, A; Rasmussen, JB. (1998) Predicting the identity and impact of future biological invaders: a priority for aquatic resource management. Can J Fish Aquat Sci 55:1759-1765.
31	Rixon, CAM; Duggan, IC; Bergeron, NMN; et al. (2005) Invasion risks posed by the aquarium trade and live fish markets on the Laurentian Great Lakes. Biodiversity Conserv 14:1365-1381.
32	Vanderploeg, HA; Nalepa, TF; Jude, DJ; et al. (2002) Dispersal and emerging ecological impacts of Ponto-Caspian species in the Laurentian Great Lakes. Can J Fish Aquat Sci 59(7):1209-1228.

Table B-2. (Continued)

Reference Number	References
33	Yan, ND; Girard, R; Boudreau, S. (2002) An introduced invertebrate predator (*Bythotrephes*) reduces zooplankton species richness. Ecol Lett 5:481-485.
34	International Association for Great Lakes Research. (2002) Research and management priorities for aquatic invasive species in the Great Lakes. Available online at http://www.iaglr.org/scipolicy/ais/ais_iaglr02.pdf
35	CANOE News. (2005) Invasive fish species found in Lake Champlain, say experts from Quebec, Vermont. Online at http://cnews.canoe.ca [11/21/05]
36	Madsen, J. (1999) A quantitative approach to predict potential nonindigenous aquatic plant species problems. ANS Update Fall/Winter 1999. Available online at http://www.glc.org/ans/ansupdate/pdf/ansdec99.pdf [11/22/05]
37	Cherwaty, S; Shear, H. (2002) Great Lakes indicators target aquatic invasive species. ANS Update Fall/Winter 2002. Available online at http://www.glc.org/ans/ansupdate/pdf/ansv8n4.pdf [11/22/05]
38	Sturtevant, R. (2002) The Great Lakes waterflea report. ANS Update Spring/Summer 2002. Available online at http://www.glc.org/ans/ansupdate/pdf/ansv8n2.pdf [11/22/05]
39	Froese, R; Pauly, D, Eds. (2005) FishBase.World Wide Web electronic publication. www.fishbase.org, version (10/2005). [12/07/05]
40	Rendall, J. (2003) Watercraft inspections: an opportunity to prevent the spread of aquatic nuisance species. ANS Update Winter/Spring 2003. Available online at http://www.glc.org/ans/ansupdate/pdf/Update-05-03.pdf [11/22/05]
41	Maki, K; Galatowitsch, S. (2003) Mail-order sales of aquatic plants: a pathway for ANS introduction. ANS Update Summer 2003. Available online at http://www.glc.org/ans/ansupdate/pdf/Update08-03.pdf [11/22/05]
42	Higbee, E; Glassner-Shwayder, K. (2004) The live food fish industry: new challenges in preventing the introduction and spread of aquatic invasive species. ANS Update Fall/Winter 2004. Available online at http://www.glc.org/ans/ansupdate/pdf/2004/ANSUpdateFW.pdf [11/22/05]
43	Rasmussen, J. (2002) The Cal-Sag and Chicago sanitary and ship canal: a perspective on the spread and control of selected aquatic nuisance fish species. US Fish and Wildlife Service, East Lansing Ecological Services Field Office, Great Lakes Coastal Program. Online at http://www.fws.gov/midwest/LaCrosseFisheries/reports/Connecting_Channels_Paper_Final.pdf [11/30/05]
44	Canadian Wildlife Federation. (2003) Invasive Species in Canada. Availablc online at http://www.cwf-fcf.org/invasive/chooseSC.asp [12/05/05]
45	Whelan, G. (2004) Largemouth bass virus continues to spread in Michigan waters. Michigan Department of Natural Resources, Lansing, MI. Available online at http://www.michigan.gov/dnr/0,1607,7-153-10364_10950-93547--,00.html [12/07/05]
46	Indiana Department of Natural Resources. (2005) Invasive species. Available online at http://www.in.gov/dnr/invasivespecies/ [12/09/05]
47	Klein, B. (2004) Making A List: Prevention Strategies for Invasive Plants in the Great Lakes States. Environmental Law Institute, Washington D.C. Available online at http://www.elistore.org [11/23/05]

Table B-2. (Cotinued)

Reference Number	References
48	Minnesota Department of Natural Resources. (2005) Invasive species. Available online at http://www.dnr.state.mn.us/invasives/index.html [12/09/05]
49	Invasive Species Specialist Group Global Invasive Species Programme. (2005) ISSG Global Invasive Species Database. Available online at http://www.issg.org/database [01/2006]
50	Berg, D. (1992) The spiny water flea, Bythotrephes cederstroemi Another unwelcome newcomer to the Great Lakes. Ohio Sea Grant College Program, Fact Sheet 49. Available online at http://sgnis.org/publicat/papers/bergdj92.pdf [01/18/06]
51	Stepien, CA; Tumeo, MA. (2006) Invasion genetics of Ponto-Caspian gobies in the Great Lakes: a 'cryptic' species, absence of founder effects, and comparative risk analysis. Biol Invasions 8:61-78.
52	USGS (2005) NAS - Nonindigenous Aquatic Species. Available online at http://nas.er.usgs.gov/ [01/17/06]
53	Bailey, SA; Reid, DF; Colautti, RI; et al. (2005) Management of nonindigenous species in the Great Lakes. Toledo J Great Lakes' Law, Science & Policy 5:101-112.
54	Pimentel, D. (2005) Aquatic nuisance species in the New York State Canal and Hudson River Systems and the Great Lakes Basin: An economic and environmental assessment. Environ Manage 35(5):692-701.
55	Dextrase, AJ; Mandrak, NE. (2006) Impacts of alien invasive species on freshwater fauna at risk in Canada. Biol Invasions 8:13-24.
56	Wisconsin Department of Natural Resources. (2005) Aquatic Invasive Species Program. Available online at http://www.dnr.wi.gov/invasives/aquatic.htm [12/09/05]

APPENDIX C. GARP MODEL VALIDATION

GARP Model Validation

The most rigorous evaluation of any model is a test of its ability to correctly predict independent data that have not been processed by the model. In this chapter, Genetic Algorithm for Rule-set Production (GARP) models were evaluated by comparing the known occurrences of three invasive species already within the Great Lakes with the predictive ability of GARP models developed for these species using occurrence data from other regions.

Model performance was assessed using area under the curve of the receiver operating characteristic curve (Sing et al., 2005) using R 2.4.0 (R Development Core Team, 2006). Area under the curve is a threshold-independent evaluation of model performance that, in this case, measures the ability of the model to differentiate between sites where a species is considered present versus where it is considered absent. Area under the curve represents the probability that, when a predicted-present site and a predicted-absent site are drawn at random, the predicted-present site will have a higher predicted value than the predicted-absent site.

Because true absence data were not available, randomly generated absence data, termed pseudo-absence data (i.e., points selected randomly from sites where the species have not been recorded as present within the Great Lakes), were used to validate the GARP models. This is a standard approach when true absence data are not available (Graham et al., 2004).

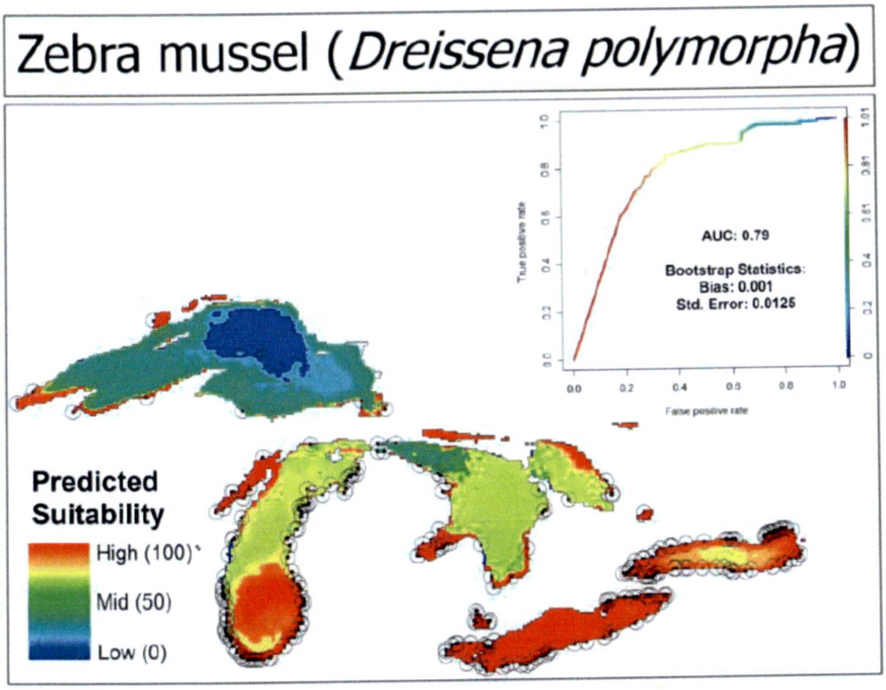

Figure C-1. GARP model validation for zebra mussel showing predicted suitability and area under the curve (inset)

GARP produces predictions of habitat suitability ranging from 0 to 100 that can be converted to a binary prediction of presence or absence by selecting a threshold. For the purpose of model validation, values above this threshold (i.e., 50) are considered present (assigned a value of 1) while values below this threshold are considered absent (assigned a value of 0). The threshold that is selected is typically the threshold that maximizes model performance which may bias estimates of model performance. Area under the curve avoids the subjectivity in the threshold selection process and, therefore, provides an unbiased evaluation of model performance by plotting the false-positive rate (i.e., over-prediction, the rate at which the model predicts the species to be present at sites at which it is considered absent) versus the true-positive rate (i.e., the rate at which the model correctly predicts known presences as present) across *all* possible thresholds. For these reasons, area under the curve is considered one of the best approaches for model validation (Pearce and Ferrier, 2000). Nonetheless, area under the curve poses three important limitations. Notable to his study are that (1) it weights over-prediction and under-prediction errors equally, (2) it does not give information about the spatial distribution of prediction errors, and (3) the size of the study area to which models are projected influences the rate of correctly predicted absences and the area under the curve scores.

Explanation of figures

The first figure, Figure C-1, is explained to help interpret the set of three figures. Figure C-1 shows the results from model validation for the zebra mussel, including a plot of the receiver operating characteristic curve with the area under the curve statistic. The colors along this curve correspond to the colors in the map of the zebra mussel predicted habitat suitability. Thus, by moving along the curve, one can stop at any color transition, say between yellow and orange (or a threshold value of 0.81 as determined by the right-hand y-axis or 81 from the legend in the map). By moving horizontally from this threshold value to the left-hand y-axis, one can determine the rate at which known presence correctly are predicted as present, also known as the true-positive rate (about 0.75 in this example). By moving vertically downward from this threshold to the x-axis, one can determine the rate at which pseudo-absences were predicted as present (the true-false rate, which is about 0.3 in this case). In other words, if values greater than 81 in the map are considered as present and values less than 81 as *absent*, we would get roughly 75% of the known occurrences correctly predicted, but it would also predict roughly 30% of presumed absences as present. In this manner, the receiver operating characteristic curve provides a means to assess the rates of false-positive predictions (predicting a species present where it is considered absent) and false-negative predictions (predicting a species absent where it is known to be present).

Model Evaluation Results

Swets (1988) suggested the following scale for determining model performance using area under the curve: 0.90–1.00 = excellent; 0.80–0.90 = good; 0.70–0.80 = fair; 0.60–0.70 = poor; ≤0.60 = fail. The area under the curve for all 3 species falls between 0.74 and 0.79, so the models would fall into the 0.70 to 0.80 category of "fair" (see Table C-1).

Table C-1. Summary of Area Under the Curve (AUC) Values and Occurrence Data Sets Used for Model Construction (Training Points) and Evaluation Points. Evaluation Points within the Great Lakes are Shown as Hollow Points on Figures C-1 to C-3

Species and common name	No. of Great Lakes evaluation points	No. and location of training points	AUC
Dreissena polymorpha zebra mussel	238	24 (Europe)	0.79
Gymnocephalus cernuus ruffe	46	183 (Europe)	0.79
Potamopyrgus antipodarum New Zealand mud snail	10	844 (Europe, Australia)	0.74

Model Evaluation

All three model validation Figures (C-1 to C-3) show (1) the predicted habitat suitability for each species within the Great Lakes when using only occurrence data from outside the Great Lakes, (2) the corresponding receiver operating characteristic curve plot and area under the curve value and bootstrap statistics for each model, and (3) the occurrence data within the Great Lakes withheld from GARP and used for evaluation of predictive performance (hollow points). Note that for reasons discussed under "Selecting Species to Model and Development of Occurrence Data" in Section 3.1, these occurrence points may not be inclusive of all known occurrences in the Great Lakes and represent only those suitable for model evaluation.

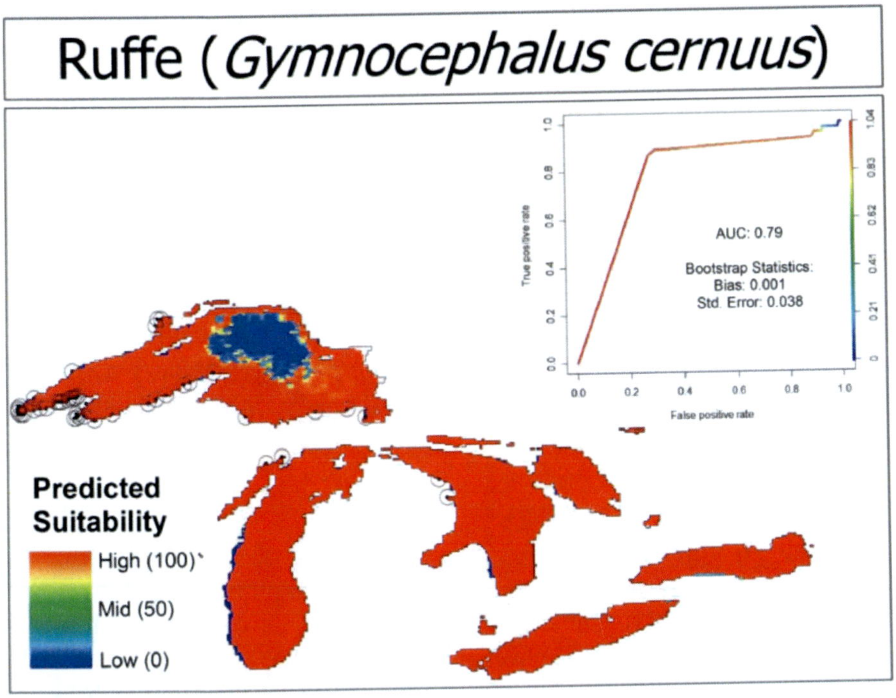

Figure C-2. GARP model validation for ruffe showing predicted suitability and area under the curve (inset)

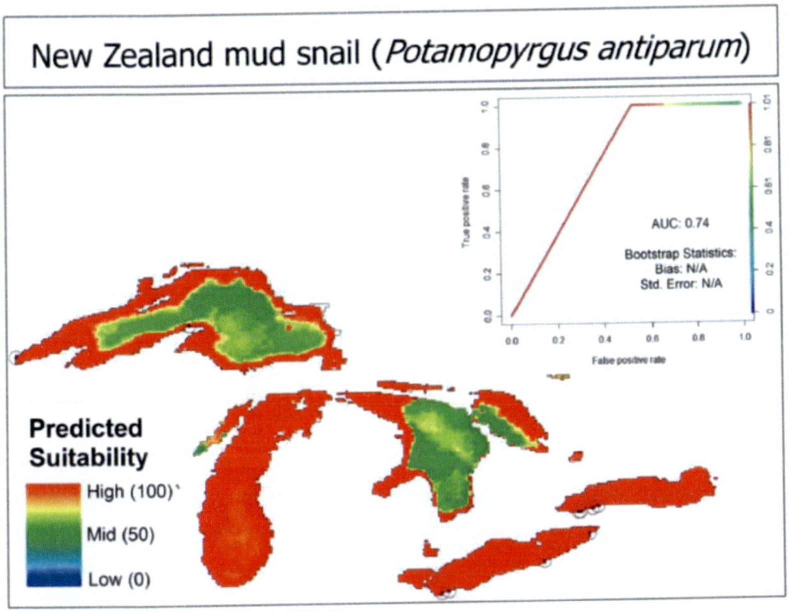

Figure C-3. GARP model validation for New Zealand mud snail showing predicted suitability and area under the curve (inset)

Taken together, these area under the curve scores and the predicted distributions suggest three important conclusions. First, in our tests, GARP models adequately predict the known distributions of potential invasive species within the Great Lakes and, therefore, may be capable of accurately identifying areas of the Great Lakes susceptible to aquatic invasive species that have yet to be introduced. Thus, distribution data from a species' existing range can produce useful predictions of invasion potential using these GARP methods. Second, the observed patterns of invasion closely match those predicted for both *known* and *potential* invaders, suggesting that Lakes Erie and Ontario, near-shore areas of all of the Great Lakes in general, Saginaw Bay in Lake Huron, Lake St. Clair (located between Lakes Erie and Huron), and Thunder Bay in Lake Superior are particularly prone to future invasion when considering environmental tolerances alone. Finally, the universally high area under the curve scores suggest that the six environmental data layers we selected as inputs for the GARP models provide useful information for predicting the potential distributions of invasive species within the Great Lakes. In sum, the model validation exercise suggests that GARP predictions provide a useful assessment of invasion potential, given the availability of adequate occurrence data outside the Great Lakes.

References for Appendix C

Graham, C. H., Ferrier, S. & Huettman, F. et al (2004). New developments in museum-based informatics and applications in biodiversity analysis. *TRENDS Ecol Evol, 19(9),* 497-503.

Pearce, J. & Ferrier, S. (2000). Evaluating the predictive performance of habitat models developed using logistic regression. *Ecol Model, 133(3),* 225–245.

R Development Core Team (2006). R: A Language and Environment for Statistical Computing. R Foundation for Statistical Computing, Vienna, Austria. Available on-line at http://www.Rproject.org.

Sing, T. Sander, O. & Beerenwinkel N. et al. (2005). ROCR: *visualizing classifier performance in R. Bioinformatics, 21,* 3940–3941.

Swets, J. A. (1988). Measuring the accuracy of diagnostic systems. *Science, 240,*1285–1293.

Appendix D. GARP Power of Prediction Analysis

Details of How GARP Power of Prediction Analysis Was Used

If environmental conditions within a particular region of the Great Lakes differ from those near a species' currently occupied range, then a prediction about the suitability of habitat cannot be made (null prediction). A prediction cannot be made because no information exists to predict whether a species under these novel conditions will or will not find suitable habitat. Unfortunately, the Genetic Algorithm for Rule-set Production (GARP), along with most species distribution models, simply predicts environments that are non-analogous to the environments found near the species' currently occupied range to be unsuitable. But, in reality, such "null" areas may actually be habitable by the species. GARP power of prediction analysis was conceived to distinguish areas that do not provide suitable habitat (predicted absence) from areas for which a prediction cannot be made (null prediction). More details about the theoretical basis for using GARP power of prediction analysis are provided in Section 3.5. The remainder of Appendix D presents the specific application of GARP power of prediction analysis to this study.

Power of prediction analyses were performed for 11 of the 14 species to help distinguish the regions for which a prediction could be made from regions where a prediction could not be made for each species. Power of prediction analysis was not performed for two of the invasive species already established in the Great Lakes, quagga mussel (*Dreissena bugensis*) and round goby (*Neogobius melanostromus*), because of a lack of occurrence data outside the Great Lakes (i.e., these species had no distribution data describing their home range, the basis for a power of prediction analysis). Also, because GARP models predict that the blueback herring (*Alosa aestivalis*) may potentially find the entire Great Lakes region as suitable habitat, no power of prediction analysis is needed, because power of prediction analyses are only concerned with areas of predicted absence.

Power of prediction analyses for species can be grouped if they share a similar home range region, eliminating the need to perform separate power of prediction analysis for each species studied. The 11 species had home-range occurrence data originating from one or more of four large regions: the northwest Pacific coast of North America, the northeast Atlantic coast of North America, the southeastern coast of Australia, and a large region comprising the northern coast of Europe and Northern Africa. Thus, we performed four power of prediction analyses.

As discussed, power of prediction analyses are only concerned with areas of predicted absence. Therefore, no power of prediction analyses were required for those species originating from southeast Australia and the Atlantic U.S. regions because the models developed for species from these regions predicted the entire Great Lakes region as suitable habitat. Similarly, no power of prediction analysis was performed for the U.S. Pacific coast region because (1) only the New Zealand mud snail, (*Potamopyrgus antipodarum)* had occurrence points within this region and (2) this species was already predicted to be fairly widespread within the Great Lakes. Thus, only Europe and North Africa remained as regions on which to focus power of prediction analyses. This complex region was subdivided into three sub-regions: the United Kingdom and northern Europe (designated UK) shoreline, the Baltic Sea, and the Ponto-Caspian Sea. GARP power of prediction analyses were performed on combinations of these regions. The gray shaded regions in Figures 7–17 and 24–27 reflect the results of the GARP power of prediction analysis.

Reverse power of prediction analysis, which processes in the opposite direction from invaded range to home range, was also performed, but these results are not included here. When we looked globally for places with conditions similar to those in the Great Lakes, we found largely salt-water environments at homologous latitudes. This result is not surprising because the vast majority of the earth's surface waters are oceanic. Assuming that salinity represents a dispersal barrier, most species found in these saline, but otherwise similar aquatic environments, would not be able to survive in the freshwater of the Great Lakes.

Grouped Results for 11 Species

Figures D1 through D3 show the results of the GARP power of prediction analyses. Multiple species are shown in most figures because the species are grouped according to the presence of species occurrence points in each sub-region. Each of the 11 species collectively shown in D1 through D3, can be associated with a corresponding GARP figure (Figures 8-17 and 31-34). Similar to the GARP predictions, the power of prediction analyses produced a continuous measure of the ability to predict by adding the binary presence/absence outcomes from the best 100 GARP models out of 1,000 total models constructed. For clarity, we simplified the figures by converting these continuous predictions into two areas: areas where prediction is possible (shown in red) and areas where prediction is not possible (shown in black). Red areas represent location where more than 50 percent of the models have predictive power and black areas depict locations in which less than 50 percent of the models could predict presence or absence.

The power of prediction analyses for the fishhook waterflea, the roach, the rudd, and the ruffe (see Figure D1) indicate that GARP can make predictions for these species everywhere throughout the Great Lakes except for a central deep-water portion of Lake Superior. GARP originally predicted a rather limited group of areas as susceptible for invasion by roach and rudd. Coupled with the power of prediction analysis for these two species, we can conclude that indeed, most of the areas thought to be low-risk areas are in fact predicted by the models as low- risk areas. Similarly, Lakes Erie and Ontario were predicted to be particularly suitable for invasion by both species, and the power of prediction analysis showed that the modeling approach had good predictive capabilities within these two lakes. Susceptibility to invasion by these two species in the upper Great Lakes was predicted to be limited primarily to near-shore areas, and the power of prediction analysis shows that many of the areas predicted to be not susceptible to invasion by these two species fall well within the red areas determined to be predictable by GARP. Thus, these, too, are valid predictions of low risk of invasion by roach and rudd. Therefore, the GARP model results, coupled with the power of prediction analysis, successfully identify large areas of the Great Lakes as definitely unsuitable for invasion by roach or rudd. These areas represent an opportunity for effectively focusing effort and resources for monitoring these species.

The power of prediction analysis for the zebra mussel and New Zealand mud snail indicate a large area within the Great Lakes that is predictable by GARP (see Figure D2 top panel). Much of the area that is indicated as predictable based on the power of prediction analysis is predicted to be highly susceptible to invasion by these two already-established NIS. The GARP models predict that the deep-water areas of Lake Superior should be unsuitable and, therefore, not at risk of future invasion by these species. However, the power of prediction analysis for these species indicates that this area is simply not predictable. Therefore, the risk to central Lake Superior from invasion by these two species is, at this point, undeterminable by GARP models alone. GARP predictions show parts of Lake Huron to be at low risk from the New Zealand mud snail and the power of prediction analysis shows that these areas can be well predicted by GARP.

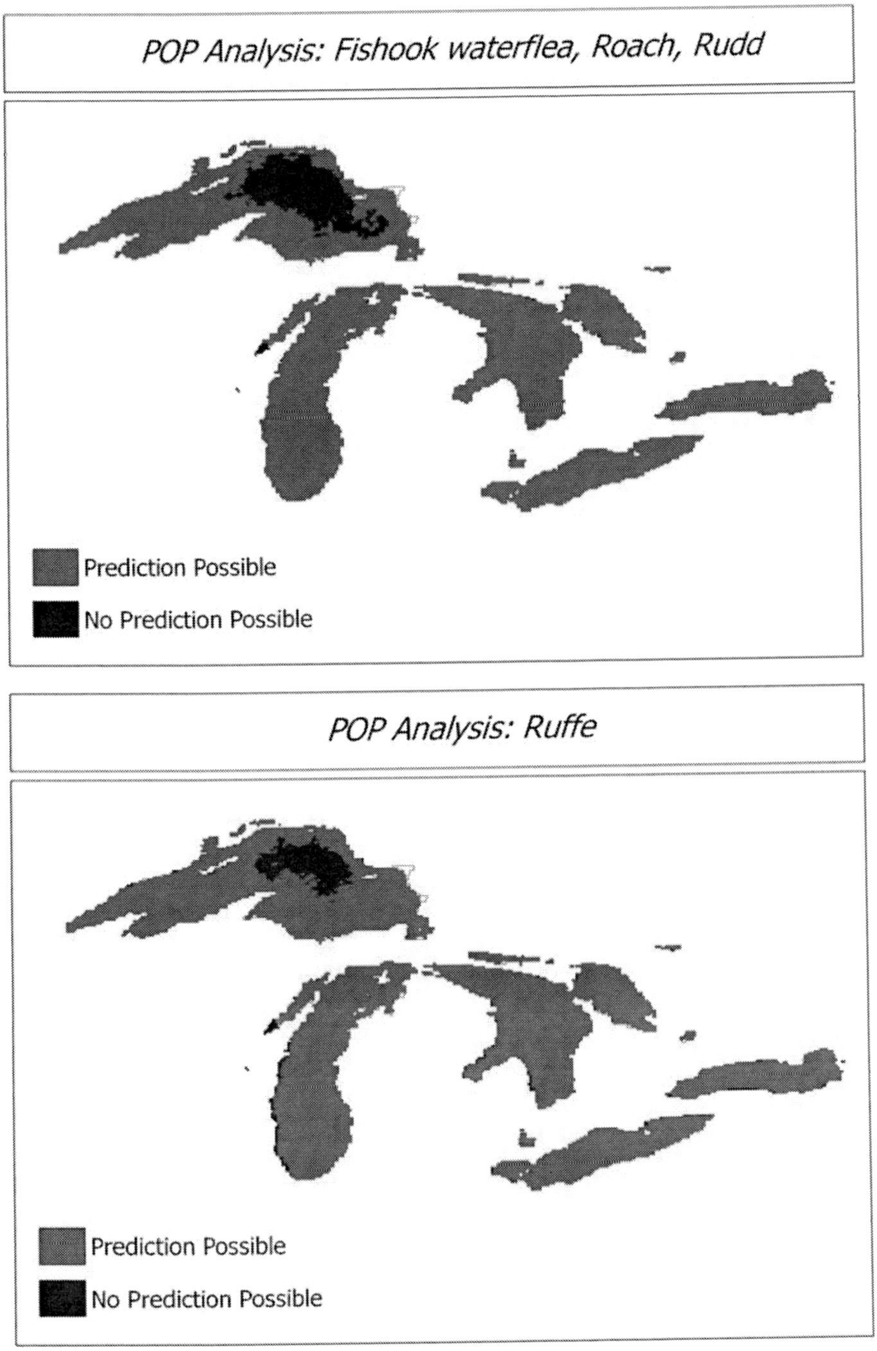

Figure D-1. Power of prediction analysis for fishhook waterflea, roach, and rudd (top) and ruffe (bottom)

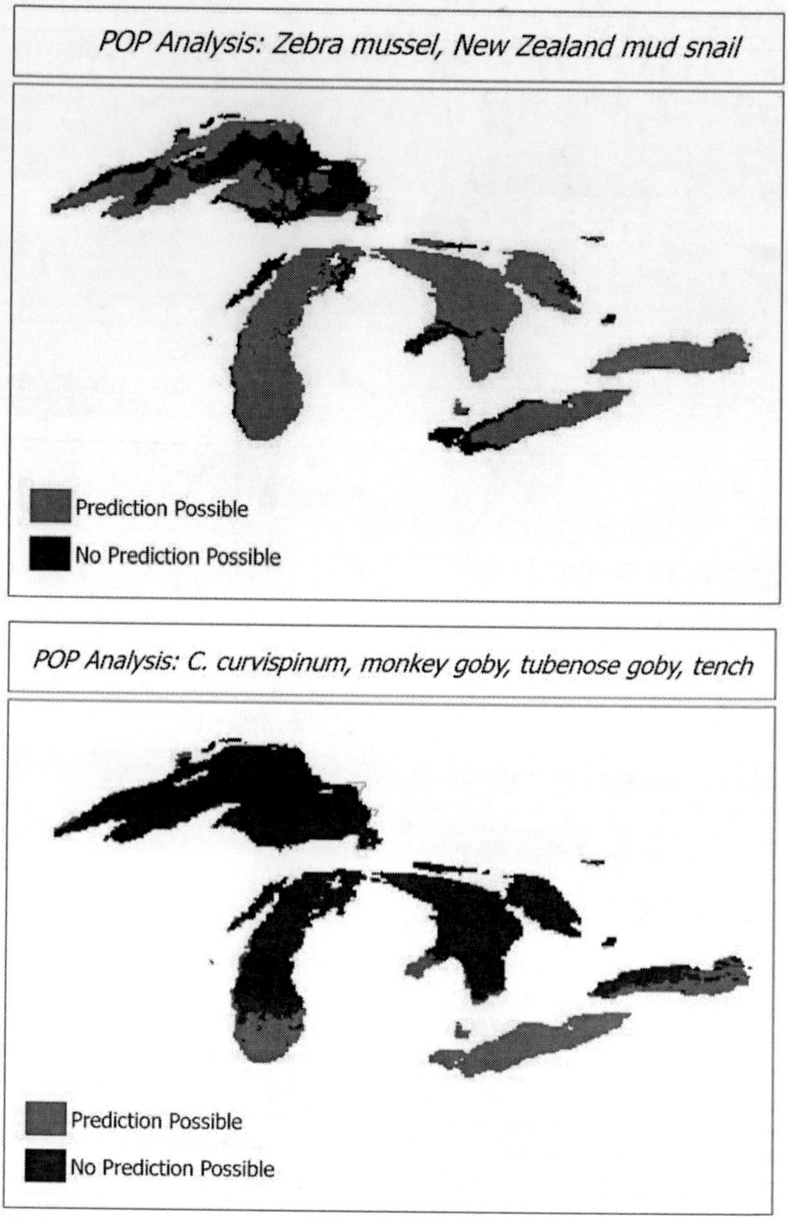

Figure D-2. Power of prediction analysis for zebra mussel and New Zealand mud snail (top) and *C. curvispinum*, monkey goby, tubenose goby and tench (bottom)

The set of power of prediction analyses that were performed for *Corophium curvispinum*, monkey goby, tubenose goby, and tench indicate that large areas within the Great Lakes are not predictable by GARP (see Figure D2 lower panel). GARP predictability zones (red areas) are restricted to Lake Erie, the southern half of Lake Ontario, and the southernmost tip of Lake Michigan. All four of these potential invasive NIS were predicted by GARP models to encounter large areas of the Great Lakes that were unsuitable as habitat, which could be classified as immune to invasion. Many of these predicted absence areas actually fall within the black zone of the power of prediction analysis, within which no GARP prediction can be made. From the GARP predictions alone, these extensive areas would have been interpreted by managers as no-invasion-risk zones. However, the areas actually are no-prediction possible zones, meaning they might be susceptible to invasion and they might not. Without the benefit of the power of prediction analysis, this condition would not have been known from the GARP predictions alone. The power of prediction analysis prevents this critical misinterpretation.

When viewed in conjunction with the corresponding GARP power of prediction analysis figures, the results present a clearer prediction of invasion-susceptible habitat areas. Without power of prediction analysis, only 5 species (the blueback herring, the sand goby, the zebra mussel, the ruffe, and the New Zealand mud snail) are predicted by GARP models to be able to find suitable habitat throughout the Great Lakes. With the power of prediction analyses, it is possible that the entire Great Lakes region could also be susceptible to three additional potential invasive species: *C. curvispinum*, the tubenose goby, and the round goby. Managers cannot rule out much of the Great Lakes as unsuitable for these three species, as was suggested by the GARP predictions alone.

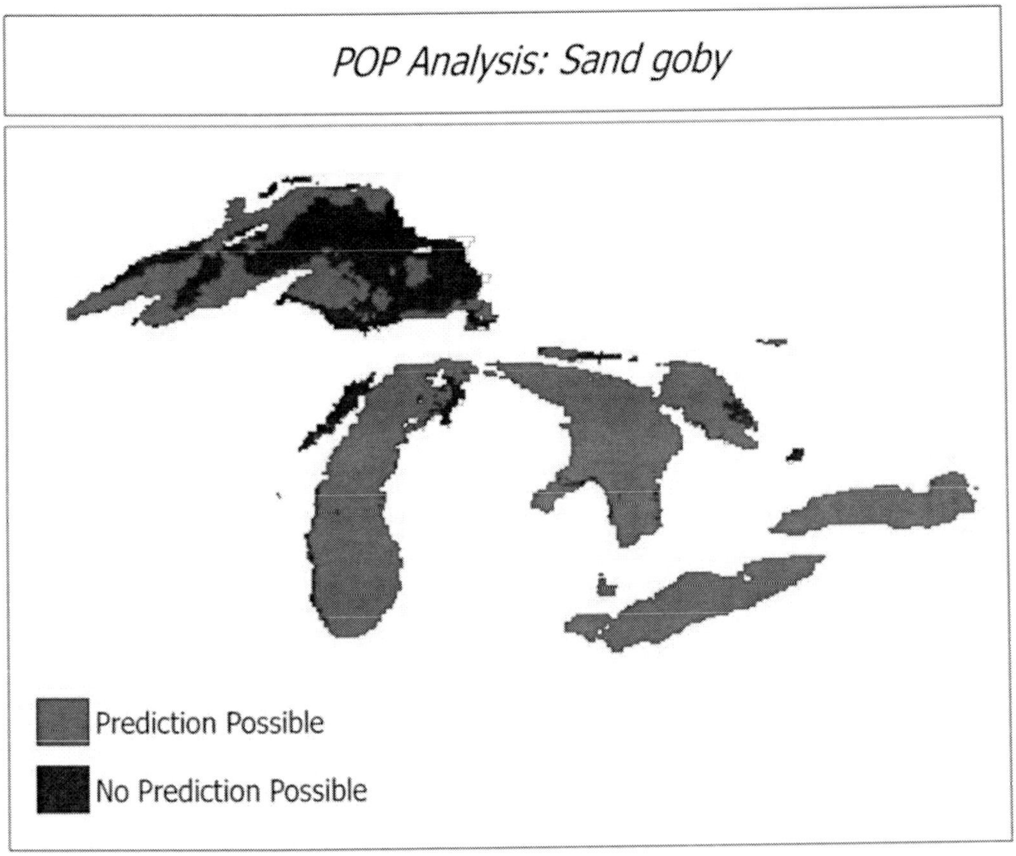

Figure D-3. Power of prediction analysis for sand goby

APPENDIX E. TABLES DEPICTING SOURCES AND DESTINATION OF BALLAST WATER DISCHARGES IN U.S. GREAT LAKES

Table E-1. Source of Ballast Water Discharges (Prior to Ballast Water Exchange) Into U.S. Great Lakes Ports (2006-2007)
Includes only discharges from vessels whose source of ballast water (prior to ballast water exchange) came from outside the Great Lakes

Source of Ballast Water	Country	Ballast Tanks Discharged	Volume Discharged (metric tons)	Vessels Discharging
Not from POC	--	79	31,323	20
Port Cartier	Canada	57	80,233	8
Antwerpen	Belgium	40	14,134	4
Baie Comeau	Canada	34	46,353	5
Port Alfred	Canada	34	18,699	2
Puerto Cabello	Venezuela	29	14,267	3
Sept Iles	Canada	28	16,014	5
Haraholmen	Sweden	22	3,586	1
Bremen	Germany	20	5,858	1
Manfredonia	Italy	18	4,364	1
St John	Canada	15	3,680	1
Ghent	Belgium	14	7,332	1
Amsterdam	Netherlands	13	3,613	1
New Haven	US	13	17,252	1
Safi	Morocco	13	7,231	1
Casablanca	Morocco	12	10,552	2
Hafnarfjordur	Iceland	12	9,718	1
Santander	Spain	12	10,633	2
Albany	US	11	1,427	1
Rotterdam	Netherlands	8	1,389	2
Hamburg	Germany	7	4,374	1
Houston	US	7	1,593	2
Thisvi	Greece	7	1,998	1
Workington	UK	7	1,477	1
Yokkaichi	Japan	7	1,512	1
Agadir	Morocco	6	3,383	1
Rouen	France	6	2,006	1
South Shields	UK	6	1,926	1
Bahia Quintero	Chile	5	16,839	1
Mostaganem	Algeria	5	1,196	1
Philadelphia	US	5	5,299	1
Porsgrunn	Norway	5	877	1
Algeciras	Spain	4	490	1
Baltimore	US	4	8	1

Table E-1. (Continued)

Source of Ballast Water	Country	Ballast Tanks Discharged	Volume Discharged (metric tons)	Vessels Discharging
Barranquilla	Colombia	4	5,098	2
Ciwandan	Indonesia	4	491	1
Dunkerque (east)	France	4	2,194	1
Hancock	US	4	9,632	1
Troy	US	4	57	4
Brake	Germany	3	970	1
Shuwaikh	Kuwait	3	739	1
Al Burayqah	Libya	2	350	1
Dammam	Saudi Arabia	2	360	1
Gros Cacouna	Canada	2	2,426	1
Holyhead	UK	2	1,374	1
Mina Jabal Ali	UAE	2	577	1
Mobile	US	2	612	1
Oulu	Finland	2	182	1
Searsport	US	2	395	1
Southampton	UK	2	548	1
Chiba Ko	Japan	1	379	1
Halifax	Canada	1	910	1
Montevideo	Uruguay	1	237	1
Port Of Pasajes	Spain	1	126	1
Santa Marta	Colombia	1	160	1
Sundsvall	Sweden	1	254	1
Thessaloniki	Greece	1	1,000	1
Veracruz	Mexico	1	719	1
Wilmington	US	1	1,500	1
TOTAL		618	381,927	107

Table E-2. Ballast Water Discharges at U. S. Great Lakes During 2006-2007 includes only vessels whose original source of ballast water (prior to ballast water exchange) came from outside the Great Lakes

US Great Lake Port	Tanks Discharged	Volume Discharged (metric tons)	Vessels Discharging
Duluth	407	184,844	58
Toledo	85	65,335	13
Superior	50	78,085	10
Green Bay	18	5,984	4
Gary	17	11,154	4
Milwaukee	11	10,768	2
Oswego	8	1,239	5
Chicago	7	17,916	3
Ludington	5	1,913	1

Table E-2. (Continued)

US Great Lake Port	Tanks Discharged	Volume Discharged (metric tons)	Vessels Discharging
Erie	4	490	1
Lorain	3	2,320	3
Menominee	2	380	2
Ashtabula	1	1,500	1
TOTAL	618	381,927	107

Table E-3. Source of 2006-2007 Ballast Water Discharges from Vessels Whose Original Source of Ballast Water (Prior to Ballast Water Exchange) Came from Outside the Great Lakes
Sorted by U.S. Great LAKE port of call

US Great Lake Port	Source of Ballast Water	Country	Tanks Discharged	Volume Discharged (metric tons)	Vessels Discharging
Ashtabula	Wilmington	US	1	1,500	1
Chicago	Bahia Quintero	Chile	5	16,839	1
Chicago	Not from POC	--	1	77	1
Chicago	Thessaloniki	Greece	1	1,000	1
Duluth	Agadir	Morocco	6	3,383	1
Duluth	Al Burayqah	Libya	2	350	1
Duluth	Albany	US	11	1,427	1
Duluth	Amsterdam	Netherlands	13	3,613	1
Duluth	Antwerpen	Belgium	33	11,668	2
Duluth	Baie Comeau	Canada	3	1,922	1
Duluth	Casablanca	Morocco	12	10,552	2
Duluth	Chiba Ko	Japan	1	379	1
Duluth	Ciwandan	Indonesia	4	491	1
Duluth	Dammam	Saudi Arabia	2	360	1
Duluth	Dunkerque (east)	France	4	2,194	1
Duluth	Ghent	Belgium	14	7,332	1
Duluth	Hamburg	Germany	7	4,374	1
Duluth	Haraholmen	Sweden	22	3,586	1
Duluth	Houston	US	7	1,593	2
Duluth	Manfredonia	Italy	18	4,364	1
Duluth	Mina Jabal Ali	UAE	2	577	1
Duluth	Mobile	US	2	612	1
Duluth	Montevideo	Uruguay	1	237	1
Duluth	Mostaganem	Algeria	5	1,196	1
Duluth	New Haven	US	13	17,252	1
Duluth	Not from POC	--	66	27,054	13

Table E-3. (Continued)

US Great Lake Port	Source of Ballast Water	Country	Tanks Discharged	Volume Discharged (metric tons)	Vessels Discharging
Duluth	Oulu	Finland	2	182	1
Duluth	Philadelphia	US	5	5,299	1
Duluth	Port Alfred	Canada	27	11,679	1
Duluth	Port Cartier	Canada	13	15,803	2
Duluth	Puerto Cabello	Venezuela	26	11,719	2
Duluth	Rotterdam	Netherlands	6	1,249	1
Duluth	Safi	Morocco	13	7,231	1
Duluth	Santa Marta	Colombia	1	160	1
Duluth	Santander	Spain	12	10,633	2
Duluth	Searsport	US	2	395	1
Duluth	Sept Iles	Canada	11	6,024	2
Duluth	Shuwaikh	Kuwait	3	739	1
Duluth	Southampton	UK	2	548	1
Duluth	St John	Canada	15	3,680	1
Duluth	Thisvi	Greece	7	1,998	1
Duluth	Workington	UK	7	1,477	1
Duluth	Yokkaichi	Japan	7	1,512	1
Erie	Algeciras	Spain	4	490	1
Gary	Baie Comeau	Canada	9	10,077	1
Gary	Not from POC	--	1	60	1
Gary	Porsgrunn	Norway	5	877	1
Gary	Rotterdam	Netherlands	2	140	1
Green Bay	Antwerpen	Belgium	3	1,082	1
Green Bay	Brake	Germany	3	970	1
Green Bay	Rouen	France	6	2,006	1
Green Bay	South Shields	UK	6	1,926	1
Lorain	Barranquilla	Colombia	1	780	1
Lorain	Not from POC	--	1	821	1
Lorain	Veracruz	Mexico	1	719	1
Ludington	Sept Iles	Canada	5	1,913	1
Menominee	Port Of Pasajes	Spain	1	126	1
Menominee	Sundsvall	Sweden	1	254	1
Milwaukee	Baie Comeau	Canada	9	10,077	1
Milwaukee	Not from POC	--	2	691	1
Oswego	Not from POC	--	4	1,182	1
Oswego	Troy	US	4	57	4
Superior	Baie Comeau	Canada	8	16,988	1
Superior	Barranquilla	Colombia	3	4,318	1
Superior	Hancock	US	4	9,632	1
Superior	Not from POC	--	1	95	1

Table E-3. (Continued)

US Great Lake Port	Source of Ballast Water	Country	Tanks Discharged	Volume Discharged (metric tons)	Vessels Discharging
Superior	Port Cartier	Canada	28	42,855	4
Superior	Puerto Cabello	Venezuela	3	2,548	1
Superior	Sept Iles	Canada	3	1,649	1
Toledo	Antwerpen	Belgium	4	1,384	1
Toledo	Baie Comeau	Canada	5	7,290	1
Toledo	Baltimore	US	4	8	1
Toledo	Bremen	Germany	20	5,858	1
Toledo	Gros Cacouna	Canada	2	2,426	1
Toledo	Hafnarfjordur	Iceland	12	9,718	1
Toledo	Halifax	Canada	1	910	1
Toledo	Holyhead	UK	2	1,374	1
Toledo	Not from POC	--	3	1,343	1
Toledo	Port Alfred	Canada	7	7,020	1
Toledo	Port Cartier	Canada	16	21,576	2
Toledo	Sept Iles	Canada	9	6,428	1
TOTAL			618	381,927	107

Table E-4. Number of Vessel-Trips Including Specified Port as a Last Five Ports of Call, for NOBOB-RM Vessels (2006)
NOBOB-RM vessels are vessels that entered the seaway without ballast on board, picked up ballast water in the great lakes, and then discharged the ballast water along with residual material at a Great Lake port. NOBOB-RM vessels must also have visited a port outside the Great Lakes during one of the last five ports of call

Port outside St. Lawrence Seaway	Country	NOBOB L5POC
Sept Iles	Canada	145
Port Cartier	Canada	77
Ijmuiden	Netherlands	23
Baie Comeau	Canada	23
Pointe Noire	Canada	22
Halifax	Canada	12
Europa Point	Gibraltar	10
Belledune	Canada	10
Antwerpen	Belgium	9
Casablanca	Morocco	9
Riga	Latvia	8
Santos	Brazil	8
Rotterdam	Netherlands	7
Stephenville	Canada	7
Amsterdam	Netherlands	7

Table E-4. (Continued)

Port outside St. Lawrence Seaway	Country	NOBOB L5POC
Ashdod	Israel	7
Police	Poland	6
Praia Mole	Brazil	6
Tynemouth	UK	5
Dunkerque (east)	France	5
Tunis	Tunisia	5
Ghent	Belgium	5
Gdansk	Poland	5
Swinoujscie	Poland	5
Oran	Algeria	5
Syros Island	Greece	5
Puerto Quetzal	Guatemala	4
Magdalen	Canada	4
Sankt-peterburg	Russia	4
Kao-hsiung	Taiwan	3
Vitoria	Brazil	3
Charlottetown	Canada	3
Hamburg	Germany	3
La Goulette	Tunisia	3
Richards Bay	South Africa	3
Coatzacoalcos	Mexico	2
Balboa	Panama	2
Corinto	Nicaragua	2
Piombino	Italy	2
Barranquilla	Colombia	2
Bahia San Nicolas	Peru	2
Arkhangelsk	Russia	2
Puerto Cabello	Venezuela	2
Rocky Point	Jamaica	2
Rouen	France	2
Tarragona	Spain	2
Puntarenas	Costa Rica	2
Summerside	Canada	2
Porto De Maceio	Brazil	2
Paranagua	Brazil	2
Santander	Spain	2
Shanghai	China	1
Chi-lung	Taiwan	1
Setubal	Portugal	1
Bunbury	Australia	1
Singapore	Singapore	1

Table E-4. (Continued)

Port outside St. Lawrence Seaway	Country	NOBOB L5POC
Cartagena	Spain	1
Campana	Argentina	1
Ceuta	Spain	1
Brunsbuttel	Germany	1
Sao Francisco	Brazil	1
Sluiskil	Netherlands	1
Fos	France	1
Szczecin	Poland	1
Terneuzen	Netherlands	1
Avonmouth	UK	1
Trieste	Italy	1
Venezia	Italy	1
Argentia	Canada	1
Aratu	Brazil	1
Xingang	China	1
Annaba	Algeria	1
Bremen	Germany	1
Recife	Brazil	1
Port Alfred	Canada	1
Port St Joe	USA	1
Point Tupper	Canada	1
Yosu-bando	Republic of Korea	1
Porto Empedocle	Italy	1
Map Ta Phut	Thailand	1
Liverpool	UK	1
Las Palmas	Spain	1
La Plata	Argentina	1
Durban	South Africa	1
Kingston Upon Hull	UK	1
Djen-djen	Algeria	1
Itea	Greece	1
Reka Luga	Russia	1
Huelva	Spain	1
Hong Kong	Hong Kong	1
Guayaquil	Ecuador	1
Safi	Morocco	1
Geraldton	Australia	1
San Lorenzo	Argentina	1
Port Of Spain	Trinidad and Tobago	1
Santa Marta	Colombia	1
Pusan	Republic of Korea	1
TOTAL		544

Table E-5. Frequency and Volume of Ballast Tank Discharges into U.S. Great Lakes Ports of Call from NOBOB-RM Vessels (2006)
NOBOB-RM vessels are vessels that entered the seaway without ballast on board, picked up ballast water in the Great Lakes, and then discharged the ballast water along with residual material at a Great Lake port. NOBOB-RM vessels must also have visited a port outside the Great Lakes during one of the last five ports of call

U.S. Great Lake port	Ballast Tanks Discharged	Volume Discharged (metric tons)
Toledo	353	511,181
Superior	340	524,446
Ashtabula	297	579,785
Duluth	289	379,999
Sandusky	116	243,137
Milwaukee	89	68,522
Gary	60	52,987
Chicago	48	76,607
Conneaut	43	76,335
Buffalo	37	73,197
Calumet	30	33,974
Lorain	13	25,729
Two Harbors	8	13,444
Brevort	7	20,914
TOTAL	1730	2,680,255

CHAPTER SOURCES

The following chapters have been previously published:

Chapter 1 – This is an edited, excerpted and augmented edition of a United States Congressional Research Service publication, Report Order Code RL32344, dated September 2, 2009.

Chapter 2 – This is an edited, excerpted and augmented edition of a United States Environmental Protection Agency publication, EPA/600/R-08-066F, dated November 2008.

INDEX

D

E

F